A. D. SAKHAROV
COLLECTED SCIENTIFIC WORKS

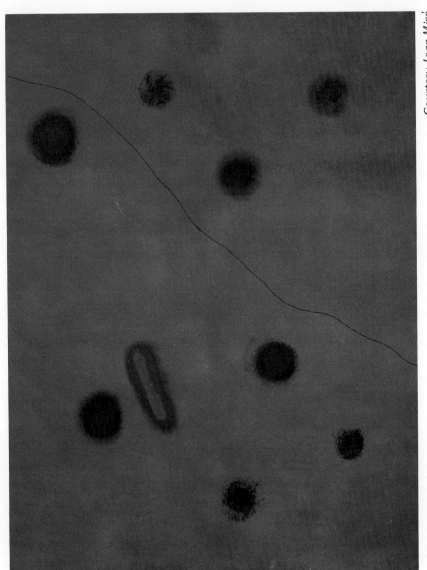

Courtesy Joan Miró

Bleu I, 4 mars 1961, Huile sur toile, 270 x 355 cm.

АКАДЕМИК
АНДРЕЙ ДМИТРИЕВИЧ
САХАРОВ

ИЗБРАННЫЕ
НАУЧНЫЕ
ТРУДЫ

ACADEMICIAN
ANDREI DMITRIEVICH SAKHAROV

COLLECTED SCIENTIFIC WORKS

Editors:

D. ter Haar, *Oxford University*
D. V. Chudnovsky, *Columbia University*
G. V. Chudnovsky, *Columbia University*

MARCEL DEKKER, INC.   New York and Basel

Library of Congress Cataloging in Publication Data
Sakharov, Andrei, 1921–
  Collected scientific works.

  Translated from Russian.
  1. Plasma (Ionized gases)—Collected works.
2. Cosmology—Collected works. 3. Particles
(Nuclear physics)—Collected works. 4. Field theory
(Physics)—Collected works. I. Title.
QC717.8.S2413 1982   539   82-13967
ISBN 0-8247-1714-7

COPYRIGHT © 1982 by MARCEL DEKKER, INC. ALL RIGHTS RESERVED
Neither this book nor any part may be reproduced or transmitted
in any form or by any means, electronic or mechanical, including
photocopying, microfilming, and recording, or by any information
storage and retrieval system, without permission in writing from
the publisher.
MARCEL DEKKER, INC.
270 Madison Avenue, New York, New York 10016
Current printing (last digit):
10 9 8 7 6 5 4 3 2 1
PRINTED IN THE UNITED STATES OF AMERICA

Preface

The collection contains 24 papers, i.e., the majority of my publications from 1947 through 1980. The summaries I have given include a brief indication of the contents and my present estimation of all my published and some unpublished papers. In particular, I have discussed my unpublished works, which I have called divertissements.* These are problems in physics and mathematics undertaken as a domestic pastime, some of them very difficult and not completely solved.

As can be seen from the summaries, I regard my main achievements to be:

1. The cycle of papers in which I collaborated with I. E. Tamm on controlled thermonuclear reactions
2. The proposal of magnetic cumulation
3. The hypothesis concerning the origin of the baryon asymmetry of the universe and the nonconservation of the baryon charge†
4. The hypothesis of the reversal of the arrow of time and the multi-sheet cosmological model of the Universe with the finite cosmological constant

---

*_Divertissements_ has been used to render Sakharov's Любительские задачи, which might be translated "pastime problems." These divertissements are contained in Part 4.—TRANS.

†My starting point was the hypothesis of cosmological _CPT_ symmetry. Nonconservation of the baryon charge, and the possible role of this effect in the appearance of the baryon asymmetry of the universe, was conjectured in 1964 by S. Weinberg. In 1967, I wrote in jest an epigram to this paper:

	(literal translation)
Из эффекта С. Окубо	From Okubo's effect
При большой температуре	At high temperature
Для Вселенной сшита шуба	For the universe has been sewn a fur coat
По ее кривой фигуре	To fit her crooked figure

Okubo is the author of the earliest paper I know of on the possibility of the nonequality of partial widths in the case of charge reflection (the total widths being equal).

5. The hypotheses of the zero-point Lagrangian of the gravitational field*

6. The linear semiempirical mass formula for mesons and baryons (the first publication in collaboration with Ya. B. Zel'dovich)

Evaluating my scientific works as a whole, I feel satisfaction, which is however considerably diminished by the thought of my unrealized plans, and of slips and errors in my published papers. Because of the peculiarity of my life's path and of my scientific style, I have always lacked knowledge of some general methods, and sometimes, knowledge of current publications as well. With the passage of years, these deficiencies have become still more perceptible.

I am very grateful to the compilers of this collection of my works; they have taken on a task that is not easy.

I hope that the published material will be of interest to some readers, including scientists who are beginning their careers.

A. D. Sakharov
July 6, 1980
Gor'kii

---

*A similar idea was formulated earlier by L. Parker, who hypothesized that the source of gravitational action is a nonlinear nonrenormalizable interaction of vacuum fields.

Editors' Introduction

We present here a collection of most of Academician A. D. Sakharov's scientific papers. This volume differs from most collected papers volumes in several respects. First of all, the author has given his own comments on the papers. Secondly, a number of specialist physicists have given their thoughts on Sakharov's contributions to their own fields of special knowledge. Finally, we include a number of "divertissements"—solutions or near-solutions to physical or mathematical conundrums which the author has collected and which constitute the main part of Part 4, and the text of Sakharov's Nobel Peace Prize Lecture (Part 5).

The main papers and specialists' comments fall naturally into three groups: papers on plasma physics (Part 1), on cosmology (Part 2), and on field theory and elementary particles (Part 3).

The present conditions of A. D. Sakharov's exile made it virtually impossible for him to supervise directly each step in the preparation of his collected works. The editors accept entire responsibility for any mistranslations or other errors that might appear in the text.

We express our sincere gratitude to S. L. Adler, J. D. Bjorken, H. P. Furth, J. Iliopoulos, H. J. Lipkin, T. H. Stix, and L. Susskind for contributing comments to Sakharov's papers. We acknowledge with thanks the help of S. Drell, N. Christ, J. Chu, P. Kusch, L. Michel, H. Robbins, M. Ruderman, and J. Wheeler. We are grateful to the American Institute of Physics, Pergamon Press, and Plenum Press for granting permission to use translations of A. D. Sakharov's papers which appeared in their translation journals. We thank J. B. Barbour, T. R. Rothman, and W. H. Furry for providing the translations of the other papers. We acknowledge with pleasure the invaluable support and care of Marcel Dekker, Inc. to this project.

The editors are indebted to Sakharov's family for their invaluable help. Mrs. E. Bonner prepared Sakharov's original notes in Russian. We are grateful to T. Yankelevich, A. Semenov, and to E. Yankelevich, whose constant advice and assistance were useful in the preparation of this volume.

The book opens with a reproduction of a painting by J. Miró chosen for this purpose by the artist. This was made possible by N. de Romilly.

D. ter Haar
D. V. Chudnovsky
G. V. Chudnovsky

Contents

Preface   *iii*

Editors' Introduction   *v*

A. D. Sakharov: An Autobiographical Note   *xi*

Chronological List of Scientific Papers of A. D. Sakharov Presented in This Volume   *xv*

Part 1   PLASMA PHYSICS   1

    Commentary: A. D. Sakharov   Plasma Physics   3

    1. Reactions Produced by μ Mesons in Hydrogen (with Ya. B. Zel'dovich)   7

    2. Theory of the Magnetic Thermonuclear Reactor, Part II   11

    3. Magnetic Cumulation (with R. Z. Lyudaev, E. N. Smirnov, Yu. I. Plyushchev, A. I. Pavlovskii, V. K. Chernyshev, E. A. Feoktistova, E. L. Zharinov, and Yu. A. Zysin)   23

    4. Magnetoimplosive Generators   29

    5. Excitation Temperature in a Gas-Discharge Plasma   43

    Commentary: Harold P. Furth   Controlled Fusion Research   49

    Commentary: Thomas H. Stix   Plasma Physics   55

Part 2 COSMOLOGY 59

> *Commentary: A. D. Sakharov* Cosmological Investigations 61
>
> 6. The Initial Stage of an Expanding Universe and the Appearance of a Nonuniform Distribution of Matter 65
>
> 7. Violation of *CP* Invariance, *C* Asymmetry, and Baryon Asymmetry of the Universe 85
>
> 8. Quark-Muonic Currents and Violation of *CP* Invariance 89
>
> 9. Antiquarks in the Universe 93
>
> 10. A Multisheet Cosmological Model 105
>
> 11. The Baryonic Asymmetry of the Universe 115
>
> 12. Cosmological Models of the Universe with Reversal of Time's Arrow 131
>
> 13. Maximum Temperature of Thermal Radiation 137
>
> *Commentary: J. D. Bjorken* Elementary Particle Physics 141
>
> *Commentary: J. Iliopoulos* The Baryon Number of the Universe 147
>
> *Commentary: L. Susskind* Matter-Antimatter Asymmetry 151

Part 3 FIELD THEORY AND ELEMENTARY PARTICLES 157

> *Commentary: A. D. Sakharov* Field Theory and Elementary Particles 159
>
> 14. Vacuum Quantum Fluctuations in Curved Space and the Theory of Gravitation 167
>
> 15. Vacuum Quantum Fluctuations in Curved Space and the Theory of Gravitation 171
>
> 16. Spectral Density of Eigenvalues of the Wave Equation and Vacuum Polarization 179

CONTENTS   ix

	17.	Scalar-Tensor Theory of Gravitation	195
	18.	The Topological Structure of Elementary Charges and *CPT* Symmetry	199
	19.	The Quark Structure and Masses of Strongly Interacting Particles (with Ya. B. Zel'dovich)	205
	20.	Mass Formula for Mesons and Baryons with Allowance for Charm	223
	21.	Mass Formula for Mesons and Baryons	227
	22.	An Estimate of the Coupling Constant Between Quarks and the Gluon Field	233
	23.	Generation of the Hard Component of Cosmic Rays	239
	24.	Interaction of the Electron and the Positron in Pair Production	255
	Commentary: Stephen L. Adler Induced Gravitation	263
	Commentary: Harry L. Lipkin Elementary Particles	271
Part 4	DIVERTISSEMENTS		281
	Divertissements		283
Part 5	NOBEL PEACE PRIZE LECTURE of 1975		289
	The Nobel Prize Lecture, 1975 Peace, Progress, and Human Rights		291

A. D. Sakharov: An Autobiographical Note

I was born on May 21, 1921, in Moscow. My father was a well-known physics teacher and the author of textbooks and popular science books. My childhood was spent in a large communal apartment where most rooms were occupied by our relatives with only a few outsiders mixed in. Our home preserved the traditional atmosphere of a large and close family—respect for hard work and ability, mutual aid, love for literature and science. My father played the piano well; his favorites were Chopin, Grieg, Beethoven, and Scriabin. During the Civil War he earned a living by playing the piano in a silent movie theater. I recall with particular fondness Maria Petrovna, my grandmother and the soul of our family, who died before World War II at the age of seventy-nine. Family influences were especially strong in my case because I received my early schooling at home; I then had difficulty relating to my own age group.

After graduating from high school with honors in 1938, I enrolled in the Physics Department of Moscow University. When the war began, our classes were evacuated to Ashkhabad, where I graduated with honors in 1942. That summer I was assigned work for several weeks in Kovrov, and then I was employed on a logging operation in a remote settlement near Melekess. My first vivid impression of the life of workers and peasants dates from that difficult summer of 1942. In September I was sent to a large arms factory on the Volga, where I worked as an engineer until 1945.

I developed several inventions to improve inspection procedures at that factory. (In my university years I did not manage to engage in

Translation courtesy of Khronika Press.
This autobiographical note was written by Andrei Sakharov (copyright © 1981) for Russian readers of *Samizdat*.—TRANS.

original scientific work.) While still at the factory in 1944, I wrote several articles on theoretical physics which I sent to Moscow for review. Those first articles have never been published, but they gave me the confidence in my powers which is essential for a scientist.

In 1945 I became a graduate student at the Lebedev Institute of Physics. My adviser, the outstanding theoretical physicist Igor Tamm, who later became a member of the Academy of Sciences and a Nobel laureate, greatly influenced my career. In 1948 I was included in Tamm's research group which developed a thermonuclear weapon. I spent the next twenty years continuously working in conditions of extraordinary tension and secrecy, at first in Moscow and then in a special research center. We were all convinced of the vital importance of our work for establishing a worldwide military equilibrium, and we were attracted by its scope.

In 1950 I collaborated with Igor Tamm in some of the earliest research on controlled thermonuclear reactions. We proposed principles for the magnetic thermal isolation of plasma. I also suggested as an immediate technical objective the use of a thermonuclear reactor to produce fissionable materials as fuel for atomic power plants. Research on controlled thermonuclear reactions is now receiving priority everywhere. The tokomak system, which is under intensive study in many countries, is most closely related to our early ideas.

In 1952 I initiated experimental work on magnetoimplosive generators (devices to transform the energy of a chemical or nuclear explosion into the energy of a magnetic field). A record magnetic field of 25 million gauss was achieved during these experiments in 1964.

In 1953 I was elected a member of the U.S.S.R. Academy of Sciences. My social and political view underwent a major evolution over the fifteen years from 1953 to 1968. In particular, my role in the development of thermonuclear weapons from 1953 to 1962, and in the preparation and execution of the thermonuclear tests, led to an increased awareness of the moral problems engendered by such activities. In the late 1950s I began a campaign to halt or to limit the testing of nuclear weapons. This brought me into conflict first with Nikita Khrushchev in 1961, and then with the Minister of Medium Machine Building,[*] Efim Slavsky, in 1962. I helped to promote the 1963 Moscow Treaty Banning Nuclear Weapon Tests in the Atmosphere, in Outer Space, and Under Water. From 1964 when I spoke out on problems of biology,[†] and espe-

---

[*] The Ministry of Medium Machine Building is responsible for nuclear weapons and industry in the U.S.S.R.—TRANS.
[†] In 1964 Sakharov spoke out at the Academy of Sciences against political interference with biology and the persecution of geneticists during a debate on the election of one of Trofim Lysenko's associates.—TRANS.

cially from 1967, I have been interested in an ever-expanding circle of questions. In 1967 I joined the Committee for Lake Baikal.* My first appeals for victims of repression date from 1966–67.

The time came in 1968 for the more detailed, public, and candid statement of my views contained in the essay "Progress, Coexistence, and Intellectual Freedom."† These same ideas were echoed seven years later in the title of my Nobel lecture: "Peace, Progress, and Human Rights." I consider the themes to be of fundamental importance and closely interconnected. My 1968 essay was a turning point in my life. It quickly gained worldwide publicity. The Soviet press was silent for some time, and then began to refer to the essay very negatively. Many critics, even sympathetic ones, considered my ideas naive and impractical. But thirteen years later, it seems to me that these ideas foreshadowed important new directions in world and Soviet politics.

After 1970, the defense of human rights and of victims of political repression became my first concern. My collaboration with Valery Chalidze and Andrei Tverdokhlebov,‡ and later with Igor Shafarevich§ and Grigory Podyapolsky,** on the Moscow Human Rights Committee was one expression of that concern. (Podyapolsky's untimely death in March 1976 was a tragedy.)

After my essay was published abroad in July 1968, I was barred from secret work and excommunicated from many privileges of the Soviet establishment. The pressure on me, my family, and friends increased in 1972, but as I came to learn more about the spreading repressions, I felt obliged to speak out in defense of some victim almost daily. In recent years I have continued to speak out as well on peace and disarmament, on freedom of contacts, movement, information and opinion, against capital punishment, on protection of the environment, and on nuclear power plants.

In 1975 I was awarded the Nobel Peace Prize. This was a great honor for me as well as recognition for the entire human rights movement in the U.S.S.R. In January 1980 I was deprived of all my official Soviet awards (the order of Lenin, three times Hero of Socialist Labor,

---

*The Committee for Lake Baikal was organized to protect Lake Baikal from industrial pollution; it was apparently sponsored by or at least tolerated by the authorities.—TRANS.

†Sakharov's essay "Progress, Coexistence and Intellectual Freedom" was first published in English by the *New York Times* and has been republished in *Sakharov Speaks* (Alfred Knopf, New York, 1974); an autobiographical note written by Sakharov in 1973 was published as an introduction to that volume.—TRANS.

‡Valery Chalidze and Andrei Tverdohklebov are physicists who have left the U.S.S.R. under pressure from the authorities.—TRANS.

§Igor Shafarevich is a Moscow mathematician who is a corresponding member of the Academy of Sciences.—TRANS.

**Grigory Podyapolsky was a geophysicist. TRANS.

the Lenin Prize, the State Prize) and banished to Gor'kii, where I am virtually isolated and watched day and night by a policeman at my door. The regime's action lacks any legal basis. It is one more example of the intensified political repression gripping our country in recent years.

Since the summer of 1969 I have been a senior scientist at the Academy of Sciences' Institute of Physics. My current scientific interests are elementary particles, gravitation, and cosmology.

I am not a professional politician. Perhaps that is why I am always bothered by questions concerning the usefulness and eventual results of my actions. I am inclined to believe that moral criteria together with uninhibited thought provide the only possible compass for these complex and contradictory problems. I shall refrain from specific predictions, but today as always I believe in the power of reason and the human spirit.

A. D. Sakharov
March 24, 1981
Gor'kii

Chronological List of Scientific Papers of A. D. Sakharov Presented in This Volume

Generation of the hard component of cosmic rays, *ZhETF** 17:686–697 (1947).

Interaction of the electron and the positron in pair production, *ZhETF* 18:631–635 (1948).

Excitation temperature in a gas-discharge plasma, *Izv. Akad. Nauk SSSR Ser. Fiz. 12 (No. 4)*:372–375 (1948).

Reactions produced by μ mesons in hydrogen (with Ya. B. Zel'dovich), *ZhETF* 32:947–949 (1957); *Sov. Phys. JETP* 5:775–777 (1957), trans.

Theory of the magnetic thermonuclear reactor, Part II, *Proceedings of the 1957 Geneva Conference on Peaceful Applications of Atomic Energy, Vol. I*, pp. 21–34, Pergamon Press, 1961.

The initial stage of an expanding universe and the appearance of a nonuniform distribution of matter, *ZhETF* 49:345–358 (1965); *Sov. Phys. JETP*, 22:241–249 (1966), trans.

Magnetic cumulation (with R. Z. Lyudaev, E. N. Smirnov, Yu. I. Plyushchev, A. I. Pavlovskii, V. K. Chernyshev, E. A. Feoktistova, E. L. Zharinov, and Yu. A. Zysin), *Dokl. Akad. Nauk SSSR*, 165:65–68 (1965); *Sov. Phys. Dokl.* 10, 1045–1047 (1966), trans.

The quark structure and masses of strongly interacting particles (with Ya. B. Zel'dovich), *J. Nuclear Phys. (U.S.S.R.)* 4:395–406 (1966); *Sov. J. Nuclear Phys.* 4:283–290 (1967), trans.

Magnetoimplosive generators, *Usp. Fiz. Nauk* 88:725–734 (1966); *Sov. Phys. Usp.* 9:294–299 (1966), trans.

Maximum temperature of thermal radiation, *ZhETF Pis'ma* 3:439–441 (1966); *JETP Letters* 3:288–289 (1966), trans.

*ZhETF = Zhurnal Eksperimental'noi i Teoreticheskoi Fiziki.

Violation of *CP* invariance, *C* asymmetry, and baryon asymmetry of the universe, *ZhETF Pis'ma* 5:32–35 (1967), *JETP Lett.* 5:24–27 (1967), trans.

Quark-muonic currents and violation of *CP* invariance, *ZhETF Pis'ma* 5:36–39 (1967); *JETP Lett.* 5:27–30 (1967), trans.

Vacuum quantum fluctuations in curved space and the theory of gravitation, *Dokl. Akad. Nauk SSSR* 177:70–71 (1967); *Sov. Phys. Dokl.* 12:1040–1041 (1968), trans.

Vacuum quantum fluctuations in curved space and the theory of gravitation. Preprint, Institute of Applied Mathematics, Moscow, 1967.

Antiquarks in the Universe, in *Problems in Theoretical Physics*, Festschrift for the 60th birthday of N. N. Bogolyubov, Nauka, Moscow, 1969, pp. 35–44.

A multisheet cosmological model. Preprint, Institute of Applied Mathematics, Moscow, 1970.

The topological structure of elementary charges and *CPT* symmetry, in *Problems of Theoretical Physics*, dedicated to the memory of I. E. Tamm, Nauka, Moscow, 1972, pp. 243–247.

Scalar-tensor theory of gravitation, *ZhETF Pis'ma* 20:189–191 (1974); *JETP Lett.* 20:81–82, (1974), trans. Erratum: *ZhETF Pis'ma* 20:692 (1974); *JETP Lett.* 20:317 (1974), trans.

Mass formula for mesons and baryons with allowance for charm, *ZhETF Pis'ma* 21:547–557 (1975); *JETP Lett.* 21:258–259 (1975), trans.

Spectral density of eigenvalues of the wave equation and vacuum polarization, *Teoret. Matemat. Fis.* 23:178–190 (1975); *Theor. Math. Phys.* 23:435–444 (1976), trans.

The baryonic asymmetry of the universe, *ZhETF* 76:1172–1181 (1979); *Sov. Phys. JETP* 49:594–599 (1979), trans.

Mass formula for mesons and baryons, *ZhETF* 78:2112–2115 (1980); *Sov. Phys. JETP* 51:1059–1060 (1980), trans.

An estimate of the coupling constant between quarks and the gluon field, *ZhETF* 79:350–353 (1980); *Sov. Phys. JETP* 52:175–176 (1980), trans.

Cosmological models of the universe with reversal of time's arrow, *ZhETF* 79:689–693 (1980); *Sov. Phys. JETP* 52:349–351 (1980), trans.

A. D. SAKHAROV
COLLECTED SCIENTIFIC WORKS

Part 1
PLASMA PHYSICS

Commentary: A. D. Sakharov

Plasma Physics

I. THEORY OF MUON CATALYSIS

1. P. N. Lebedev Physics Institute Report, 1948 [not included in this volume]. Having become acquainted with a paper of Frisch in which he discussed a possible alternative interpretation of the experiments of Powell, Lattes, and Occhialini (discovery of the π meson) by means of a μ-catalysis reaction, I wrote a report considering the possibility of realizing μ catalysis of a D + D reaction on a macroscopic scale with a positive energy balance, and I made some calculations. Subsequently, muon catalysis was the subject of many experiments (the first were Lederman's) and theoretical studies. It has now been found that muon catalysis with positive energy balance may occur in the D + T system (Ponomarev et al.).

2. In 1957, I published a paper together with Ya. B. Zel'dovich in the *Zhurnal Eksperimental'noĭ i Teoreticheskoĭ Fiziki* on muon catalysis ["Reactions produced by μ mesons in hydrogen"; Paper 1[†]]. In particular, we estimated the cross section of the "enticing" reaction D + μH → μD + H.

II. CONTROLLED THERMONUCLEAR REACTIONS

1. Reports during 1950–1951 (in collaboration with I. E. Tamm). Here we proposed the principle of magnetic thermalization, determined the transport coefficients of a "magnetized" plasma (thermal conductivity, diffusion, and thermal diffusion), and proposed a toroidal configuration in a stationary and a nonstationary variant; the latter is discussed in connection with the problems of plasma instability.

[†]Paper numbers refer to Sakharov's papers as they appear in this volume.

2. A report of 1951. I proposed a thermonuclear breeder in which neutrons from the thermonuclear D + T reaction are used to accumulate plutonium or uranium 233 and tritium. Plutonium and uranium 233 are burned in relatively simple (nonbreeder) reactors, producing energy, tritium, and fissile materials. Evidently it is in this direction that a controlled thermonuclear reaction can for the first time achieve practical significance. The papers of 1 and 2 were presented by I. V. Kurchatov during his visit to the Harwell Laboratory in 1956 (during the visit of Khrushchev and Bulganin to Great Britain) and were then published in the *Proceedings of the Geneva Conference on the Peaceful Uses of Atomic Energy* ["Theory of the magnetic nuclear reactor"; Paper 2].

In P. T. Astashenkov's book *Achievements of Academician Kurchatov*, which was published in 1979 and begins with an account of Kurchatov's visit to Harwell, the names of Tamm and myself are not mentioned. I. N. Golovin's biography of Kurchatov, published several years earlier, gives our names.

3. In a seminar in 1960 (perhaps 1961, I do not recall), I discussed the possibility of realizing a controlled thermonuclear reaction by means of a laser.

4. In connection with the idea of "explosive breeding" proposed by a number of authors, I have made a number of additional suggestions in various seminars, proposing in particular the use of a subterranean "corrugated" chamber. In this variant, the soil plays the part of the strong walls needed to withstand the pressure of the explosion products, and hermetic sealing is achieved by a thin-walled chamber. The entire project carries the danger of radioactive contamination, and it should perhaps be carried out on the moon, and the fuel transported to the Earth by spacecraft.

III. MAGNETIC CUMULATION

1. In 1951–1952 I proposed a method for obtaining superstrong magnetic fields using the energy of an explosion. The constructions MK-1 and MK-2 described in comment 4 below were proposed.

2. The problem of magnetic cumulation in cylindrical geometry has an exact (self-similar) solution in the case of finite conductivity of the shell if the incompressible shell implodes in accordance with the law $R \sim \sqrt{-t}$ (M. P. Shumaev and A. D. Sakharov, 1952). The solution is expressed in terms of hypergeometric functions.

3. The problem of the hydrodynamic stability of imploding cavities of an incompressible ideal fluid (exact solution). For high harmonics, there is conservation of the adiabatic invariant E/ω (E is the energy and ω the frequency). This last result also applies to compressible ideal fluids.

4. "Magnetoimplosive generators" [Paper 4]. A review of the experimental and theoretical studies on magnetic cumulation during the years 1951–1966.

Paper 1

Reactions Produced by µ Mesons in Hydrogen

At present there is evidence that at Berkeley [1] in a bubble chamber filled with liquid hydrogen having a varying deuterium content, there was observed a nuclear reaction catalyzed by µ mesons. The possibility of such a reaction was first pointed out by Frank [2] in connection with the analysis of π-μ disintegrations in emulsions. This process was investigated in liquid deuterium by both of the authors of this article, independently of one another [3,4].

The presence of a µ meson changes the form of the potential barrier, which previously prevented nuclear reactions among slow proton and deuteron nuclei, increasing sharply the penetrability of the barrier and making possible the reactions

$p + D = He^3$

$D + D = He^3 + n$

$D + D = T + p$

In the presence of tritium there are also possible the reactions

$D + T = He^4 + n$

$T + T = He^4 + 2n$

$p + T = He^4$

The reaction $p + p = D + e^+ + \nu$, catalyzed by mesons, is practically impossible, since in addition to the barrier there is also the factor of a low probability for the beta process.

It has been predicted [4] that the probability of the reaction in flight is low, the production of mesomolecules practically always leads to

Coauthor: Ya. B. Zel'dovich.
Source: Мю-мезонный катализ, ЖЭТФ 32:947-949 (1957) (совместно с Я. Б. Зельдовичем). Reprinted from *Sov. Phys. JETP* 5:775–777 (1957), translated by D. A. Kellog, with permission from American Institute of Physics.

nuclear reactions, the rate of the process is determined by the production of mesomolecules, and the probability of mesomolecule formation during the lifetime of a meson may amount to several hundredths or even tenths, depending on the arrangement of the mesomolecule levels.

The experimental data of Alvarez [1] show that in natural hydrogen (deuterium content ratio 1:7000), the reaction $p + d = \text{He}^3$ occurs on the average once for each 150 mesons. If the deuterium ratio is 1:300, the reaction occurs once per 40 mesons, and if the ratio is 1:20, the reaction occurs once per 33 mesons. Furthermore, the energy of the resulting He^3 (5.4 MeV) is carried away by the μ meson, so that monochromatic μ mesons are observed while the reaction is taking place. The relatively high probability found for the reaction in the natural mixture is explained [1] by the transfer of the meson from the proton to the deuteron (charge exchange): $p\mu + d = p + d\mu$. Because of the difference in reduced mass, the energy of the $D\mu$ bond (2655 eV) is greater by $\Delta E = 135$ eV than the energy of the $p\mu$ bond. Therefore the charge-exchange process appears to be irreversible under the experimental conditions.

We shall give a rough estimate of the probability of the transition. If ΔE is equal to zero, the cross section should be of the order of πa^2, where a is the radius of the Bohr orbit of the mesoatom, 2.5×10^{-11} cm.

Indeed, if the masses of the two nuclei are equal, with $\Delta E = 0$, the states of the systems Σ_g and Σ_u appear to be proper, and the cross section for charge exchange can be expressed by the scattering lengths a_g and a_u of these states in a continuous spectrum: $\sigma = \pi(a_g - a_u)^2$. When $\Delta E \neq 0$, but is still small with respect to the molecular dissociation energy, then

$$\sigma = \pi(a_g - a_u)^2 \frac{v_f}{v_i}$$

where v_i is the velocity before collision and v_f is the velocity after collision.

In actual fact, ΔE is of the same order as the dissociation energy, so that the formula is corrected at least in order of magnitude. If v_i is small, $\sigma \sim 1/v_i$. It follows that in order of magnitude,

$$\sigma \approx \pi a^2 \frac{v_*}{v_i}$$

where v_* is the characteristic velocity corresponding to 1.35 eV. Using the masses of the proton and deuteron, $v_* = 2 \times 10^7$ cm/sec. Such an estimate yields qualitative agreement with the observed facts. Calculation shows that saturation should be reached at deuterium concentrations of 1 in 300 to 1 in 20. In the natural mixture of hydrogen, the probability of Dμ production and consequently the probability of the reaction should be threefold less than in enriched mixtures; experimentally it is found to be four to five times less.

Let us examine the reaction in the pDμ molecule. The observed high probability a of the giving up of meson energy does not agree with the hypothesis that the reaction proceeds like an electric dipole transition $E1$, since a becomes 2×10^{-3} for the meson under this hypothesis. Therefore, to estimate the probability we cannot use the experimental cross section for $p + D = \text{He}^3 + \gamma$ (as was done before [4]), since under the conditions of the measurement, it is precisely the cross section of the $E1$ process which is observed.

In the case of zero orbital momentum, the system $p + D$ can be found in either of the states $+3/2$ or $+1/2$. The transition to He^3 ($+1/2$ state) is possible in the first case as $M1$ and $E2$, and in the second case as $M1$ and $E0$.[†]

The conversion coefficients have the following values: for $M1$, $a = 10^{-4}$; for $E2$, $a = 0.1$; for $E0$, only the giving up of meson energy is possible (the probability of pair production e^+ and e^- at the expense of $E0$ is 10^{-3} of the probability of the giving up of meson energy in the case of $p + D$, but is of the order of unity in the case of $p + \text{H}^3 = \text{He}^4$).

Calculations concerning barrier penetration for the pDμ molecule under adiabatic investigation of the motion of the proton and deuteron yield $\psi^2(0) = 6 \times 10^{-27}$ cm^{-3}.

For the mirror reaction $n + D$, it is assumed that for the purposes of computation [8] that the process proceeds from the state $3/2$ as the result of $M1$. Experimentally for thermal neutrons [9], $\sigma = 5.7 \times 10^{-28}$ cm^2 with $v = 2200$ m/sec and $\sigma v = 1.3 \times 10^{-22}$ cm^3/sec.

Hence for mesomolecules the probability of the reaction ($\tau = 2.15 \times 10^{-6}$ being the meson lifetime) is

$$w = \frac{\sigma v \psi^2(0)}{1/\tau + \sigma v \psi^2(0)} = 0.6$$

During the approach of the proton and deuteron in the spin state $1/2$, the approximate determination of the magnitude of the monopole moment was carried out by examining one charged particle with the wave function $\psi = (1/\sqrt{2\pi\lambda}\ r)\ e^{-r/\lambda}$ in the final (combined) state and $\psi = \psi(0)(1 - \lambda/r)$ in the initial state [continuous spectrum, $\psi(0)$ being the

---
[†] Church and Weneser [7] have recently drawn attention to the role of $E0$ in the case of internal conversion in the transition $J \rightarrow J \neq 0$.

previously calculated wave function under the barrier, $\psi^2(0) = 6 \times 10^{-27}$ cm^{-3}].

The probability of a process with a release of energy, to a meson in a $p + D$, in spin state ½, and with $\lambda = 2.4 \times 10^{-13}$ cm, turned out to be equal to 0.5.[†]

Thus from the rough estimates made above, it follows that the probability observed by Alvarez for the process which involves release of energy to the meson, and the probability of the process which involves emission of a gamma quantum, can both be close to unity during the meson lifetime.

In a more accurate investigation, not only would it be necessary to take into account the fact that the process is not adiabatic (thus involving terms of the order of the meson-nucleon mass ratio), but it would also be necessary to make a separate investigation of the nuclear reaction with different values of total molecular spin.

Note Added in Proof (February 9, 1957)

The probabilities for mesomolecule production in the collisions $D\mu + p = Dp\mu$ and $D\mu + D = D_2\mu$ differ not only because of the different positions of the excited vibrational levels of the molecules [4], but also because in $pD\mu$ the center of mass does not coincide with the charge center, and thus possesses a dipole moment $\frac{1}{3}ea$. Therefore in the collision of slow $D\mu + p$ there is possible the dipole transition $E1$ in the molecule into the momentum state 1, with energy given up to the electron. In the case of $D\mu + D$, only the $E2$ transition into momentum state 2 competes with the $E0$ transition investigated [4].

REFERENCES

1. L. W. Alvarez et al., Lithographed document, December 1956.
2. F. C. Frank, *Nature 160*:525 (1947).
3. A. D. Sakharov, *Report of the Physics Institute*, Academy of Sciences, USSR (1948).
4. Ia. B. Zel'dovich, *Dokl. Akad. Nauk SSSR 95*:454 (1954).
5. G. M. Griffiths and I. B. Warren, *Proc. Phys. Soc. 68*:781 (1955).
6. D. H. Wilkinson, *Phil. Mag. 43*:659 (1952).
7. E. L. Church and J. Weneser, *Phys. Rev. 103*:1035 (1956).
8. N. Austern, *Phys. Rev. 83*:672 (1951); *85*:147 (1952).
9. Kaplan, Ringo, and Wilzbach, *Phys. Rev. 87*:785 (1952).
10. E. E. Salpeter, *Phys. Rev. 88*:547 (1952).

[†]Choosing $\lambda = \hbar/\sqrt{2ME}$, where M is the reduced mass of p and D, and E is the binding energy 5.4 MeV.

Paper 2

Theory of the Magnetic Thermonuclear Reactor, Part II[†]

The properties of a high-temperature plasma in a magnetic field were discussed in a paper by Tamm [1], in which he demonstrated the possibility of the realization of a magnetic thermonuclear reactor (MTR). In this paper we shall consider other questions concerning the theory of MTRs.

  I. *Thermonuclear reactions. Bremsstrahlung*
 II. *Calculation of the large model. Critical radius. Local phenomena near the wall*
III. *Power of magnetization. Optimal construction. Performance of active matter*
IV. *Drift in a nonuniform magnetic field. Suspended current. Inductive stabilization*
 V. *Problem of plasma instability*

I. THERMONUCLEAR REACTIONS. BREMSSTRAHLUNG

The following reactions may take place in an MTR:

$$D^2 + D^2 \rightarrow H^3 + p + \underline{3\ MeV} + \underline{1\ MeV}$$
$$D^2 + D^2 \rightarrow He^3 + n + \underline{0.82\ MeV} + 2.46\ MeV$$ primary reactions

$$He^3 + D^2 \rightarrow He^4 + p + \underline{14.5\ MeV} + \underline{3.7\ MeV}$$
$$H^3 + D^2 \rightarrow He^4 + n + 14\ MeV + \underline{3.5\ MeV}$$ secondary reactions

Energies supplied by the charged particles and which maintain the

Source: Теория магнитного термоядерного реактора, часть II, в сборнике Физика плазмы и проблема управляемых термоядерных реакций, т. 1, Изд-во АН СССР, 1958. Reprinted from *Proceedings 1957 Geneva Conference on Peaceful Applications of Atomic Energy*, Vol. 1, pp. 21–34, 1961, Pergamon Press, with permission from Pergamon Press Ltd.
[†]Work done in 1951; Parts I and III were written by I. E. Tamm.

thermonuclear reaction in the MTR are underlined. The time τ required for a collision which results in a reaction is given by

$$\begin{aligned} \tau_D^{-1} &= N_D([\sigma v]_1 + [\sigma v]_2) \\ \tau_{H^4}^{-1} &= N_D[\sigma v]_{H^3+D} \\ \tau_{He^3}^{-1} &+ N_D[\sigma v]_{He^3+D} \end{aligned} \qquad (1.1)$$

where N_D is the number of deuterons per cubic centimeter and $[\sigma v]$ is the Maxwellian distribution of the cross section for relative velocity.

Table 1 gives, to indicate the order of magnitude of quantities, the times τ corresponding to $N_D = 0.77 \times 10^{14}$ cm^{-3} (using 1950 data).

Figure 1 of the Appendix shows, as a temperature function, the amount of energy Q_1 per cubic centimeter per second supplied by charged particles at the same density $N_D = 0.77 \times 10^{14}$ (only primary reactions are included[†]).

The energy Q_2, radiated by *Bremsstrahlung* (x-rays), and calculated by the formula

$$Q_2 = aN_D^2 \sqrt{T_{MeV}} (1 + \alpha T) \qquad (1.2)$$

is given for comparison.

Here $\alpha = 5.05$ MeV^{-1} (this term results primarily from electron-electron collisions), and $a = 1.7 \times 10^{-22}$. Figure 1 of the Appendix shows that a self-sustaining reaction $Q_1 > Q_2$ is possible in a system of large dimensions.

II. CALCULATION OF THE LARGE MODEL. CRITICAL RADIUS. PHENOMENA NEAR THE WALLS[‡]

Let us consider a straight infinite cylinder of radius R. We disregard effects which depend on neutral particles. With this assumption there is no radial ion current j present (in the stationary case).

According to Ref. 1,

$$j \sim \frac{n^2}{T^{1/2}}\left(\frac{\nabla n}{n} + \frac{1}{4}\frac{\nabla T}{T}\right) \qquad (2.1)$$

from which we have for $j = 0$,

$$nT^{1/4} = \text{const} = n_0 T_0^{1/4} \qquad (2.1a)$$

[†] $Q_1 = \dfrac{N_D}{\tau_D} \dfrac{3(MeV) + 1(MeV) + 0.82(MeV)}{4} 1.6 \times 10^{-6}.$

[‡] An account of a large model (disregarding boundary effect and secondary reactions) was first given by Tamm in October 1950. Results which include the effects of processes with neutral particles were obtained by him [1] for a small-density system (temperature jump).

Table 1

T,keV	Lifetime of D with respect to both reactions (primary)	Lifetime of H^3	Lifetime of He^3
10	7620	134	27,400
20	1770	34	8850
50	421	15.3	322
100	202	13.8	89.9
200	112	18.0	51.2
300	85.5	23.0	46.4

where n_0 and T_0 are the number of deuterons in 1 cm^3 and the temperature on the axis of cylinder, respectively. From (2.1a) it can be seen that the pressure is maximal in the center and falls off to a small value at the edge. According to Ref. 1, the magnetic field in a system with cylindrical symmetry varies in such a way that the sum of the gas pressure and magnetic field pressure

$$\frac{H^2}{8\pi} + 2nT = \text{const} \tag{2.1b}$$

Further, we write the equation for the conservation of energy. Let π, in erg/cm^2 sec, be the heat flow, due to the thermal conductivity of the plasma. We have

$$\nabla \pi = \left(\frac{n}{N_D}\right)^2 Q(T) \tag{2.1c}$$

where $Q = Q_1 - Q_2 = 0.77 \times 10^{14}$ erg/cm^3 sec. Then, for T in ergs,

$$\pi = \frac{3.6 \times 10^{-8} n^2}{H^2 \sqrt{T}} \left[\nabla T + \frac{7}{2(\sqrt{M/2m} + 1/\sqrt{2} + 41/8)} \frac{T}{n} \nabla n \right] \tag{2.2}$$

From (2.1a) we have $(T/n)\nabla n = -\frac{1}{4}\nabla T$, i.e.,

$$\pi = -\frac{3.5 \times 10^{-8} n^2}{H^2 \sqrt{T}} \nabla T \tag{2.1d}$$

By examining Eqs. (2.1a) and (2.1d) together, it is possible to find the distribution of all quantities as a function of the tube radius as well as the critical value of the radius at which the release of nuclear energy is equal to the heat lost to the walls. Dimensionless variables were introduced for discussion purposes. The magnetic field appears in the combination HR; from this it follows that the critical radius R_c is inversely proportional to H_0.[†] Neglecting secondary reactions and boundary effects, it was found that for $T_0 = 107$ keV,

$$R_c H_0 \approx 10^7 \text{ G cm} \qquad (2.3)$$

For $H_0 = 25{,}000$ G, we have $R_c = 400$ cm.

We note that the numerical value of the coefficient of thermal diffusion, equal to $\tfrac{1}{4}$ and appearing in front of $\nabla T/T$ in formula (2.1), has fundamental importance for the realization of MTR. This can be seen easily from the following considerations. The order of magnitude of the heat emission of any body is proportional to (in the stationary case, when the emission of energy is not particularly inhomogeneous)

$$I = L \int_{T_1}^{T_0} \varkappa(T)\, dT$$

where L is the linear dimension, $\varkappa$ the thermal conductivity, T_0 the temperature at the center, and T_1 the temperature at the boundary. In our case $\varkappa \sim n^2/T^{1/2}$. Let $n \sim T^{-\alpha}$, where α, in general, is not necessarily equal to $\tfrac{1}{4}$ [generalization of formula (2.1a)].

We have $\varkappa \sim 1/T^{1/2 + 2\alpha}$. For $\alpha = \tfrac{1}{4}$, $I \sim \ln(T_0/T_1) \approx 15$; when $\alpha > \tfrac{1}{4}$, I can be rather large. For example, for $\alpha = \tfrac{1}{2}$, $I \sim (T_0/T_1)^{1/2} \approx 10^3$, while for $\alpha < \tfrac{1}{4}$, the integral converges for $T_1 \to 0$. It has been proved that the presence of additives in deuterium plasma having $Z > 1$ (helium, air, etc.) decreases α, i.e., it acts in a favorable direction (calculations by Fradkin).

In Figs. 2 and 3 (in the Appendix) the distributions of temperature T and particle density n for the case of $H_0 = 25{,}000$ G are shown; T_1 is taken as $1000°$ (see below). The magnetic field in the center falls off to a small value and therefore n_0 can be computed from the formula

$$n_0 = \frac{H_0^2}{16\pi T} \qquad (2.4)$$

For $H_0 = 25{,}000$ G and $T = 100$ keV $= 1.6 \times 10^{-7}$ ergs, we have $n_0 = 0.77 \times 10^{14}$ cm^{-3}. Figure 4 (in the Appendix) shows the energy release per unit of volume with n_0 given by formula (2.4) and $H_0 = 25{,}000$ G. Obviously, it is most reasonable to operate at the lowest temperature at which the thermonuclear reaction is self-sustaining.

At the present time calculations do not take account of secondary reactions and of the increase of *Bremsstrahlung* if He^3 and He^4 appear in the system. True, the first of the factors mentioned (lowering R_c) is

---

[†]This result can be easily understood without any calculations. Heat release per centimeter of cylinder $\sim R^2 Q n_0^2$. Heat loss is proportional to $\sim n_0^2 T^{1/2} H_0^{-2}$. In order to find R_c we equate these expressions. Then $R_c^2 \sim T Q^{-1} H_0^{-2}$.

more important. Evidently one can additionally decrease R_c by forced burning of He^3. One can obtain only very preliminary qualitative estimates of the method of solution in the vicinity of the wall, where an important role is played by the neutral atoms and molecules which come from the wall. (A similar problem was solved by Tamm for the case of a small model, with the occurrence of a temperature jump.)

We disregard both recombination in the plasma volume (we will consider only a recombination at the walls) and collision of neutral atoms with one another.

With the above assumptions one can find a qualitative diagram of the solution (Fig. 1) and also check the validity of the starting assumptions. The ion temperature jumps to T' in the vicinity of the wall. Slow neutral particles leaving the wall undergo charge transfer at a very small distance from the wall (of the order of 1 mm). Fast neutral particles emerge with ranges of the order of 1 cm. The difference in the ranges of fast and slow neutral particles is due to the fact that the mean free path is equal to the product of the time of free flight and the particle velocity. The time of free flight is determined by the relative velocities of neutral and charged particles and changes with the velocity of the neutral particle by only a factor of $\sqrt{2}$ at most.

Let us discuss a limiting case, when the probability of ionization is

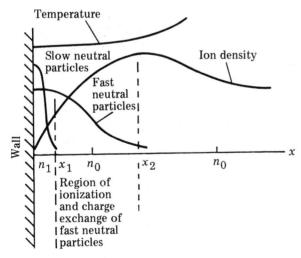

Figure 1 Region of ionization and charge exchange of slow neutral particles.

equal to the product of some α and the probability of charge exchange; $\alpha \ll 1$. Let the total current of fast neutral particles be j_n. We have $\pi_0 \sim \frac{3}{2} T' j_n + \pi_1$, where π_0 is the heat flow in the region $x > x_1$, and π_1 is the heat flow in the region $x < x_1$, which, according to the analysis below, is many times smaller and can be disregarded.

According to the theory of albedo, the probability of ionization of fast neutral particles is $1 - 1/(1 + \sqrt{\alpha}) \simeq \sqrt{\alpha}^{\dagger}$. Therefore the ion current is $j_i = \sqrt{\alpha} j_n$. The probability of ionization of slow particles is αj_n, i.e., smaller than the probability of ionization for a fast particle; it can be neglected. At $x < x_1$, n is small. Therefore $\pi_1 \sim \nabla n^2$, and the temperature can be considered as constant:

$$\pi_1 = \frac{7}{2} T' j_i = \frac{7}{2} \frac{4 \times 10^{-10} \sqrt{T'} \nabla n^2}{H^2}$$

$$\pi_1 = \frac{7}{3} \sqrt{\alpha} \pi_0$$
(2.5)

$\nabla(n^2) = n_1^2/\alpha$, where n_1' is the number of ions at point x_1.

$$x_1 = \frac{1}{\sigma n_1} \frac{v_0}{v_1}$$
(2.6)

where σ is the charge-transfer cross section, v_0 is the velocity of slow neutral particles, and v_1 is ion velocity.

At present, matching solutions in the regions $x < x_1$; $x_1 < x < x_2$, and $x > x_2$ have not been considered. We limit ourselves to a preliminary evaluation of the thermal current, for which it is possible to have a temperature jump of 10 eV (applicable to conditions of a large model; with a small model a temperature jump is certain to occur). We take $n_1 = 1.4 \times 10^{15}$ cm^{-3}, $\alpha = 1$ (i.e., we are near the limit of applicability of the above-mentioned theory), $T' = 1.6 \times 10^{-11}$ ergs, $\sigma = 3 \times 10^{-15}$ cm^{-2}, $H = 50{,}000$ G, $v_0/v_1 = 0.05$ (wall at room temperature). We obtain $x_1 = 0.01$ cm (i.e., order of magnitude of the Larmor circle for ions, and in this case we are near the limit of applicability of the theory); $\pi = 5 \times 10^8 \sim 50$ W/cm^2, which has the correct order of magnitude.

III. POWER OF MAGNETIZATION. OPTIMAL CONSTRUCTION. PERFORMANCE OF ACTIVE MATTER

The basic parameters of MTR are shown in Fig. 2. We will find the optimal relation of ∂ to d, securing a minimum mass of copper and power

---

†Ionization is similar to absorption, and charge transfer is similar to scattering. The albedo of half-space is $2/(1 + \sqrt{\alpha}) - 1$.

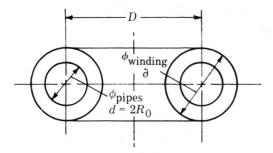

Figure 2

of magnetization in the self-sustaining region. The ratio D/∂ is obviously determined by engineering considerations and is of the order 3–5.

The product $d(\partial - d)$ is proportional to $H_0 R_0$ and should be considered as given. We are looking for a minimum of $D(\partial^2 - d^2) \sim \partial(\partial^2 - d^2)$, which can be found to occur at $\partial \approx 2.2d$. We take

$$\partial = 2d \qquad (3.1)$$
$$D = 6d$$

and the power of magnetization $P \sim H_0^2 d$. Power emitted from nuclear reaction and yield from active matter is

$$W \sim n_0^2 d^3 \sim H_0^4 d^3 \sim P^2 d$$

For characterization of numerical coefficients in these formulas we will consider the following example (in the following, we will always have this particular example in mind whenever we mention numerical parameters):

$H_0 = 50{,}000$ G
$d = 4$ m
$\partial = 8$ m
$D = 24$ m
$n_0 = 3.0 \times 10^{14}$ cm^{-3}
$T_0 = 100$ keV

Volume $= 0.96 \times 10^9$ cm^3
Area $= 0.96 \times 10^7$ cm^2
Release of thermonuclear power
17.6×10^6 erg/cm^3 sec

The weight of the copper in the winding (taking a filling factor $k = 0.5$) is 13,000 tons. The current density in the winding is 400 amp/cm^2 (average 200 amp/cm^2).

The power of magnetization is about 400,000 kW (and slightly larger when the nonuniform wrapping of the coil and other structural considerations are taken into account). Release of thermonuclear energy (assuming that, on the average, reaction takes place in half the volume of the tube with the above-mentioned speed of reaction) is

$$W = 8.8 \times 10^{15} \text{erg/sec} = 880{,}000 \text{ kW}$$

With this, the number of burned D nuclei per second is

$$\frac{8.8 \times 10^{13} \text{erg/sec}}{1.6 \times 10^{-6} \text{erg/MeV}} \frac{4}{3.3 \text{MeV} + 4 \text{MeV}} = 3 \times 10^{22} \frac{D \text{ nuclei}}{\text{sec}}$$

(which amounts to 150 g/24 hours). One can expect to obtain about 100 g/24 hours of tritium or 80 times more than U^{233}.[†] Increasing the power P and the weight of the copper by a factor of 2.5 increases this output 8.5 times (without change in current density). Increasing current density n times, we can reduce linear dimensions by a factor of $n^{1/2}$ without changing the product $H_0 R_0$. In this case the weight of copper will be reduced by a factor of $n^{3/2}$, and the power of magnetization and the yield of active substances will increase by a factor of $n^{1/2}$.

IV. DRIFT IN A NONUNIFORM MAGNETIC FIELD. SUSPENDED CURRENT. INDUCTIVE STABILIZATION

The magnetic field in a MTR (with neglected screening by plasma currents) coincides with the field of the direct current. Nonuniformity of the magnetic field leads to rather dangerous drift effects (Fig. 3). For the particle having mass M at point A the field is directed along the z-axis and the gradient of the field along the x axis, $\partial H_z/\partial x = -H_z/x$.

Suspended Current

Let us consider the motion of charged particles in the magnetic field induced by the coil of the MTR ($\sim$50,000 G) and by the current on the

---
[†]We note, however, that the energy value of U^{233} which can burn in simple reactors significantly exceeds the release of heat in a thermonuclear reactor.

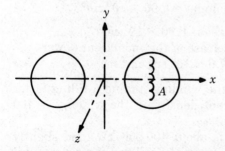

Figure 3

axis (200 A) due to a ring conductor passing along the axis of the tube (Fig. 4). In such a field magnetic lines of force have spiral shape. On the basis of the numerical illustrations above, the guiding center of the Larmor circle of a particle moving along the magnetic line of force encircles the y axis 40 times more often than the axial current (z axis). The divergence of the velocity vector field for the motion of the guiding center of Larmor circles, neglecting drift, should vanish (this follows from Liouville's theorem). On this motion drift is superimposed. The divergence of this vector field should also vanish; therefore the projection of the resulting trajectories on the cross section of the toroid (xy plane) represents closed trajectories displaced by some distance Δ from their position in the absence of drift. An estimate of the drift shows that this quantity is always sufficiently small. For example, for protons with 40 McV energy ($v_0 = 5.2 \times 10^9$ cm/sec) the spiral velocity is 2×10^7 cm/sec, and the velocity of drift is 1.5×10^6 cm/sec. From this, $\Delta \sim 20$ cm. We note that in this case we are avoiding the difficulty connected with the presence of volume charges. The question of how to produce an axial current arises. At the present time it is not clear if it is possible to put cables through the hot region which would support the axial ring and would carry current and cooling water. It is possible to form such a configuration of protective fields, for example, by means of strong current in cables, which would protect the cables from being hit by hot gas. Let us consider another possibility, i.e., suspension of the axial ring with the help of a magnetic field (an additional horizontal field with $H' \sim 100$ G will not change the qualitative picture of the magnetic field in the toroid).

The ring material should withstand high temperatures, since the only way of cooling it is by thermal radiation, corresponding, even at $T = 1400°C$, to about 40 W/cm^2 [$\pi = 5 \times 10^{-5} T^4$ (K)], i.e., a very small quantity. One of the possible ways of making a ring that operates at such high temperatures is to use metal tubes with a high melting point and containing melted light metal (Li, Be, Al, etc.).

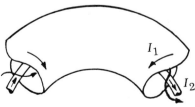

Figure 4

Direct Current—200,000 amp. The total power necessary to sustain a direct current is 2000 to 10,000 kW. Great difficulties would be encountered in the transfer of this energy (in the form of radio frequency) to the ring and in the rectification of the alternating current.

A second means of antidrift stabilization, which is technically much more feasible and which is therefore necessary to examine carefully, is the formation of an axial current directly in the plasma by the method of induction. It is not clear if, in using this method, the high-temperature plasma is not destroyed at the moment when the induction current vanishes.

V. PROBLEM OF PLASMA INSTABILITY

It is necessary to determine whether in plasma with a magnetic field disturbances exist which, according to the equations of plasma dynamics, grow in time (exponentially, or according to a power law). It is necessary to consider a series of cases. Most theoretical and experimental studies have dealt with the current flow in a plasma parallel to the external magnetic field, where turbulent instabilities of the plasma were found. One might also suspect the presence of instability in a nonuniform plasma in the presence of a drift current. At the present time this problem has merely been postulated.

APPENDIX

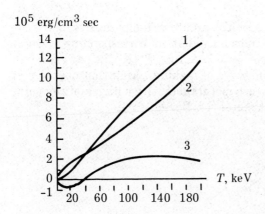

Figure A.1 1: Energy supplied by charged particles; 2: energy of *Bremsstrahlung*; 3: their difference.

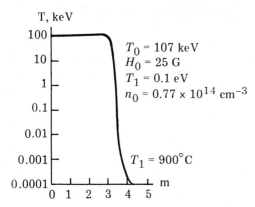

Figure A.2 Temperature distribution.

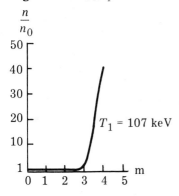

Figure A.3 Density distribution.

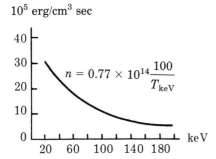

Figure A.4 Total released power.

REFERENCE

1. I. E. Tamm, in *Proceedings of the 1957 Geneva Conference on Peaceful Applications of Atomic Energy*, Vol. 1, *Plasma Physics and the Problem of Controlled Thermonuclear Reactions*, (M. A. Leontovich, ed.), Pergamon Press, New York, 1961.

Paper 3

Magnetic Cumulation

Any explosion is an abundant source of mechanical and thermal energy. In 1951 Sakharov proposed a possible way of converting this energy to magnetic form, together with the general design of devices for producing very strong fields and currents by the explosive deformation of current-carrying conductors. The process has been called magnetic cumulation. Here we describe briefly two typical generators of this type: the MK-1 (which employs compression of an axial magnetic field) and the MK-2 (ejection of the magnetic field from a solenoid and subsequent compression by the walls of a coaxial line).

Terletskii [2] published a short note on explosive compression of an axial field; a subsequent publication [1] later showed that similar experiments were done at about the same time at Los Alamos. No device resembling the MK-2 has previously been described.

I. The MK-1 is a metal tube surrounded by an explosive charge and containing an axial magnetic field. The charge produces rapid symmetrical compression and this induces a current in the tube which tends to maintain the field constant. For an ideally conducting tube

$$\Phi = \pi R^2 H = \pi R_0^2 H_0 = \text{const}$$

while the field strength and magnetic energy are inversely proportional to the square of the internal radius,

$$H = \frac{H_0 R_0^2}{R^2}$$

Coauthors: R. Z. Lyudaev, E. N. Smirnov, Yu. I. Plyushchev, A. I. Pavlovskii, V. K. Chernyshev, E. A. Feoktistova, E. I. Zharinov, and Yu. A. Zysin.
Source: Магнитная кумуляция, Доклады АН СССР 165:65-68 (1965) (соавторы: Р. З. Людаев, Е. Н. Смирнов, Ю. И. Плющев, А. И. Павловский, В. К. Чернышев, Е. А. Феоктистова, Е. И. Жаринов, Ю. А. Зысин). Reprinted from *Sov. Phys. Dokl.* 10:1045–1047 (1966), with permission from American Institute of Physics.

$$W = \frac{W_0 R_0^2}{R^2}$$

where R_0, W_0, H_0 correspond to the initial values of the internal pipe radius, the magnetic energy, and the field strength, respectively.

The magnetic flux decreases if the conductivity is finite, the decisive parameter being $\eta = (4\pi\sigma R v/c^2)^{1/2}$, where σ is the conductivity and $v = -dR/dt$. The flux is conserved if $\eta \gg 1$. We have $\eta = $ const if $v \sim 1/R$. The flux follows $\Phi \sim R^\alpha$, where $\alpha = 2.26\eta^{-1}$ for η large.

Preliminary tests with aluminum tubes about 100 mm in diameter gave fields of 10^6 Oe; in one test with a stainless steel tube the field was 25×10^6 Oe for a final diameter of 4 mm (field pressure 25×10^6 atm). Figure 1 shows the oscillogram of the field in this test. The part corresponding to $H > 25 \times 10^6$ Oe lies outside the frame.

Figure 2 shows the general design of the generator. This test employed a charge giving very rapid and fairly symmetrical compression. The initial magnetic field was produced by a coil wound with aluminum foil on the tube; this coil was energized by a bank of condensers. The initial field penetrated into the free space, although the tube was not slotted, because stainless steel has a fairly low conductivity. The internal surface was coated with 20 μm of copper to provide better trapping of the field during compression.

Alternatively, the initial field may be provided by an MK-2 (see below), which can give strong fields over very large volumes; we recorded a field of 5×10^6 Oe in a volume of 100 cm^3 in a test with a copper tube 300 mm in diameter.

These fields clearly are not the strongest possible, for any desired H

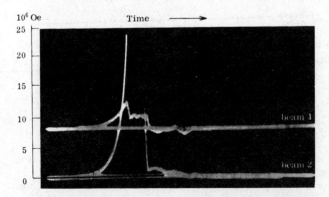

Figure 1 Field oscillograms. Beam 1: background (signal from shorted leads); beam 2: signal from probe turn 1.5 mm in diameter (with RC integration).

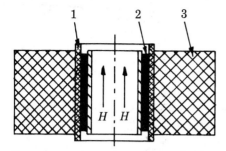

Figure 2 The MK-1. 1: Stainless-steel tube; 2: winding of aluminum foil; 3: explosive.

may, in principle, be provided by a tube compressed symmetrically at the appropriate speed, if, of course, the walls remain conducting in spite of the very large circulating currents.

II. The MK-2 consists of a central metal tube and a coaxial external spiral connected to a tube of the same diameter (vessel) attached to the inner tube (Fig. 3). The latter contains a long cylindrical charge that is detonated from one end of the spiral. Another charge may be placed around the vessel and spiral. A bank of condensers is discharged through the circuit formed by spiral, vessel, and inner tube. The explosion causes the tube to diverge as a cone, which reaches the start of the spiral at the instant when the discharge current is maximal. The subsequent propagation along the tube gives rise to a pattern analogous to that from driving a metal cone along the spiral with the detonation velocity: the point of contact with the coil moves along a helix, the number of unshorted turns decreasing, and with it the inductance of the generator. The cone eventually reaches the start of the vessel to form a coaxial volume, whose length and inductance continue to decrease. The fall in the inductance leads to increase in the current I and magnetic energy W. Rapid deformation leads to conservation of the flux, i.e., $\Phi = LI \approx L_0 I_0$, so

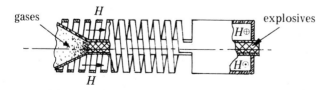

Figure 3 The MK-2.

$$I = \frac{\Phi}{L} \approx \frac{L_0}{L} I_0$$
$$W = \frac{\Phi^2}{2L} \approx \frac{L_0}{L} W_0$$

in which L_0, I_0, and W_0 are the initial values. The magnetic energy increases on account of the work done against the forces exerted by the field on the wall of the central tube.

The MK-2 has produced currents of 5×10^7 A with a final inductance of 0.01 μH; some tests have given a current amplification of over a thousandfold (to 10^8 A or more). The MK-2 has produced fields of 1 to 1.5×10^6 Oe in volumes of several liters, the stored energy being 1 to 2×10^7 J, or 10–20% of that produced by the explosive within the vessel (the flux is conserved during the deformation of the coaxial part).

III. A load may be coupled to the MK-2 either directly (in which case the load inductance is the final one and must be kept small) or via a transformer (inductive coupling of the load to the generator).

A transformer allows one to transfer a substantial fraction of the explosion energy to the load; for example, 50% of the magnetic energy has been extracted from a small generator. This allows one to accumulate considerable energy in a load of substantial inductance and to place the load at some distance where it may be shielded from the explosion. In addition, a multistage generator may be based on this principle: the initial magnetic energy in the first generator is provided by a permanent magnet, and the energy produced in it is transferred via a transformer to a second generator, where it is further amplified and transferred to a third, and so on.

Another method of transferring the energy from the generator to the load is by interruption of the current-carrying circuit by means of an additional explosive charge, the magnetic flux being ejected from the MK-2 into the load (rupture method). This method allows one to transfer over 50% of the energy generated by the MK-2 to a load consisting of inductance and resistance. The transfer time has been 0.5×10^{-6} sec in some trials.

IV. These generators provide the basis for relatively small single-shot accelerators to produce 100–1000 BeV, as well as for plasma studies, acceleration of dense bodies to speeds around 1000 km/sec (production of stellar temperatures and pressures in the laboratory), shock-wave generation, research on the equation of state and properties of materials, and examination of effects on meteorite impact on spacecraft.

A generator of this type is under development for iron-free betatrons [3], and trials have been made of remote supply of the electromagnet. Tests have also been done on coaxial electrodynamic accelerators; a velocity of 100 km/sec has been recorded for the metal vapor from aluminum foil of initial mass about 2 g.

These generators provide access to the very strong fields needed in the study of effects such as magnetoresistance in metals and semiconductors, magnetooptics, and so on.

REFERENCES

1. C. M. Fowler, W. B. Garn, and R. S. Caird, *J. Appl. Phys. 31*:588 (1960).
2. Ya. P. Terletskii, *ZhETF 32*:387 (1957); *Sov. Phys. JETP 5*:301 (1957), trans.
3. A. I. Pavlovskii, G. D. Kuleshov, G. V. Sklizkov, Yu. A. Zysin, and A. I. Gerasimov, *DAN 160 1*:68 (1965); *Sov. Phys. Dokl. 10*:30 (1965), trans.

Paper 4

Magnetoimplosive Generators

Several recent experimental and theoretical papers are devoted to the use of explosions to produce ultrastrong magnetic fields [1-9]. The same topic was also the subject of a recent international conference (Rome, September 1965). In the U.S.A. and in the U.S.S.R., fields of 15-25 million gauss were attained in individual experiments. Somewhat weaker fields (2-5 million gauss) can be attained relatively simply. Prospects are being uncovered for investigating electric, optical, and elastic properties of different substances in such hitherto unattainable magnetic fields. In addition, the magnetoimplosive generators can be used to feed pulsed accelerators of charged particles and also for some other purposes (plasma physics research, launching of bodies, etc.).

In this article we describe the physical and structural principles of magnetoimplosive generators and their characteristics, and touch upon problems involving their application.

We use the term *magnetocumulative generators* (MC generators), which agrees with Soviet usage and reflects the basic phenomenon that occurs in these systems, namely, the compression (cumulation) of the magnetic flux.

I. PRINCIPLE OF MC

In spring 1952 R. Z. Lyudaev, E. A. Feoktistova, G. A. Tsyrkov, and A. A. Chvileva realized the first implosion experiment in the U.S.S.R. aimed at obtaining superstrong magnetic fields. The experimental setup is shown in Fig. 1 (generators of this type have been designated

Source: Взрывомагнитные генераторы, УФН 88:725-734 (1966). Reprinted from *Sov. Phys. Uspekhi* 9:294-299 (1966), translated by J. G. Adashko, with permission from American Institute of Physics.

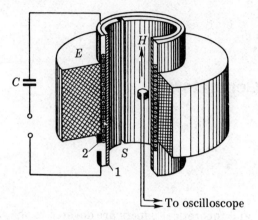

Figure 1 Diagram of MK-1 generator.

MK-1). A longitudinal magnetic field is produced inside a hollow metallic cylinder 1 by discharging capacitor bank C through the solenoid winding 2 (to ensure rapid penetration of the field inside the cylinder, a narrow oblique slot S is cut in the cylinder; subsequently this slot is closed by collapse). An explosive charge E is placed outside the cylinder. A converging cylindrical shock wave is excited in this charge (it is set off either by an electric multiple-point initiation system, or by special detonation "lenses"). The instant of initiation is chosen such that the compression of the cylinder begins when the current in the solenoid winding is a maximum.

When the cylinder moves under the influence of the detonation wave with a velocity exceeding 1 km/sec, the entire process of contraction is so rapid that in first approximation one can neglect the ohmic losses in the cylinder and regard the cylinder as an ideal conductor (for refinements see below). The electric field in an ideal conductor is equal to zero, i.e., the magnetic flux $\Phi = \pi R^2 H$ enclosed in the contracting cylindrical cavity does not change during the contraction. The magnetic field increases in this ideal case in proportion to $1/R^2$, and the energy of the magnetic field, which is equal to $W = (H^2/8\pi)\pi R^2 l$, where l is the length of the cylinder, increases in the same ratio:

$$\Phi = \pi R^2 H = \text{const}$$
$$H = \frac{H_0 R_0^2}{R^2} \tag{1}$$
$$W = \frac{W_0 R_0^2}{R^2}$$

$H \to \infty$ and $W \to \infty$ as $R \to 0$

Of course, in reality the magnetic flux decreases and infinite values of H and W are unattainable. The flux in experiments of this type is usually lower by a factor 2–3, as was the case in 1952. In addition, at a certain value of R the motion of the cylinder is stopped by the counterpressure of the magnetic field. Nonetheless, even in first experiments a field of one million gauss was attained (at an initial field of 30 kG). Measurement of the fields was by means of an induction pickup. A more detailed discussion of systems of the MK-1 type will be presented later; for the present let us consider the operation of magnetocumulative generators from the electrical engineering point of view.

We can state that the MC generators are based essentially on the same principle used in all other devices in which mechanical energy is converted into electricity. Let us consider a circuit whose inductance L can vary under the influence of external forces. At first we neglect the resistance of the circuit. We have ($\sim$ stands for proportionality)

$$\Phi = LI = \text{const}$$

$$I \sim \frac{1}{L} \tag{2}$$

$$W = \frac{LI^2}{2} = \frac{\Phi^2}{2L} \approx \frac{L_0 W_0}{L}$$

i.e., the magnetic field energy W increases with decreasing inductance. In the presence of resistance R in the deformed circuit we obtain in lieu of (2),

$$\Phi = \Phi_0 e^{-\int (R/L)dt}$$

$$W = \frac{W_0 L_0}{L} e^{-2\int (R/L)dt} \tag{3}$$

Many systems were proposed in the U.S.S.R. and in other countries, besides the systems MK-1, in which the energy and intensity of the magnetic field increases when current-carrying circuits are compressed by implosion products. A typical variant is the MK-2 generator, which we now proceed to describe.

II. MK-2

Figures 2 and 3 show a photograph and the diagram of an MC generator (MK-2) which is of particular interest for the production of strong currents (up to 10^8 A) and very large magnetic field energies (with up to 20% of the explosive energy converted into magnetic field energy at

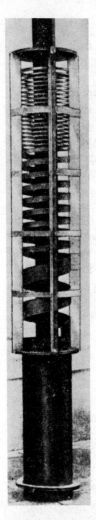

Figure 2 Photograph of the MK-2 generator.

relatively high values of the field intensity, up to 2×10^6 Oe). The operating principle of the MK-2 reduces to the following (Figs. 4a−c). When the capsule DC is detonated, the detonation wave propagates through an explosive charge located inside metallic tube 1. The tube stretches, forming at each given instant a cone (Fig. 4a) which short-circuits first the helix 2 and then the solid shell 3 in such a way that the inductance of the circuit, made up of the helix 2, tube 1, and the coaxial

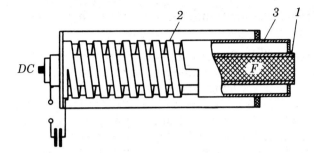

Figure 3 Diagram of the MK-2 generator.

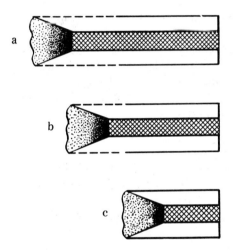

Figure 4 Stages of operation of the MK-2 generator.

section 3, decreases continuously as the detonation wave propagates to the right. The expanding tube compresses the magnetic field at the same time, increasing its energy. During the last stage of operation of the MC generator (Fig. 4c), the helix is completely disconnected and the magnetic field, which coincides in this case with the DC field, is now compressed in the decreasing volume between the external and internal walls of the coaxial.

Practical realization of MK-2 systems with high output characteristics has called for extensive research on the part of a large staff, concluded essentially in 1956 (the first MK-2 generator was constructed in 1952, and in 1953 currents up to 100 million amperes were obtained).

During the course of these investigations, the following problems were solved:

1. Prevention of cracks in the expanding tube, high symmetry of expansion, choice of the material (copper was used in all but the largest systems), choice of dimensions and tolerances, assurance of homogeneity of the explosives, etc. The results were monitored during this stage with the aid of pulsed x-ray photography and other methods.
2. Minimization of magnetic flux losses in the helical section—use of helices with variable pitch and with continuous transition to the coaxial section, beveling the edges of the helix turns at the flare angle of the tube, etc.
3. Design and experimental development of the coaxial section (arbitrarily called the "beaker"), in which the main growth of the magnetic field energy takes place.
4. Development of systems for transformer coupling the electric energy from the coaxial section (Fig. 5). This makes it possible to connect the MK-2 generator to loads with much larger inductances.
5. Development of systems in which the initial magnetic flux is produced with the aid of permanent magnets.
6. Development of cascade systems coupled by transformers.

The MK-2 generators were developed in the U.S.S.R. in a large assortment of sizes and power ratings. With the generator weighing 150 kg (the weight of the explosive in the beaker is 15 kg), more than 10^7 J goes over into magnetic field energy. A capacitor bank with the same discharge energy would be a very complicated and expensive structure.

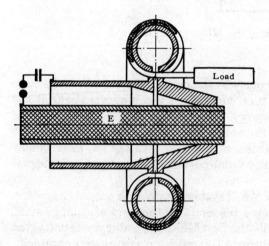

Figure 5 MC transformer system.

III. THEORY OF MK-1 AND PRODUCTION OF ULTRASTRONG FIELDS

The simplest from the theoretical point of view is the MK-1 system. In addition, this is precisely the system yielding the strongest magnetic fields (up to 25 million oersted). Let us consider, using MK-1 as an example, some principal features of the operation of MC generators, and let us make allowance first for the finite conductivity of the cylinder that compresses the magnetic flux. Actually the conductivity of a metal depends on its density and temperature, which vary during the course of the contraction. If this dependence, as well as the equation of state of the cylinder material, is assumed known, then simultaneous solution of the system of partial differential equations describing one-dimensional radial contraction and the flow of induced eddy currents in the moving metal can be achieved by numerical methods (the numerical solution of partial differential equations involves reducing them to difference equations and can be carried out in practice only with high-speed computers). Of course, no analytic solution can be obtained for the general case. However, the role of the finite conductivity of the cylinder material can be explained qualitatively with the aid of the approximate skin-layer method known from electrical engineering. In this method the exact distribution of the current in the cylinder is approximated by an "exponential profile" (4). The current density is

$$j(r) = j_0(t)e^{-x/\delta(t)} \tag{4}$$

Here $\delta(t)$ is the depth of the skin layer and $x = -R(t)$, where $R(t)$ is the radius of the cylindrical cavity; this function is assumed known. The relation (4) is exact if H increases on the boundary exponentially: $H \sim \exp(\lambda t)$, where λ is a constant. In this case δ is constant and is equal to

$$\delta = \sqrt{\frac{1}{4\pi \times 10^{-9} \sigma \lambda}} \tag{5}$$

where σ is the conductivity in ohm^{-1} cm^{-1}, δ is in cm, and λ in sec^{-1}. In the skin-layer method, formulas (4) and (5) are used for an arbitrary law of increase of $H(t)$, with the quantity λ in (5) assumed equal to $(1/H)(dH/dt)$. Regarding the change in the flux as a correction, we get $\lambda = -2(1/R)(dR/dt)$. In addition, it is obvious that $H = 0.4\pi j_0 \delta$, and the electric field, in a system moving together with the metal, is $E = (10^{-8}/2\pi R)(d\Phi/dt) = j_0/\sigma$. Combining all these formulas we get

$$\frac{1}{\Phi}\frac{d\Phi}{dt} = \frac{\alpha}{R}\frac{dR}{dt} \tag{6}$$

where the "loss coefficient" α is equal to (putting $v = |dR/dt|$)

$$\alpha = \frac{10^4}{(0.2\pi)^{1/2}\sigma^{1/2}R^{1/2}v^{1/2}} \tag{7}$$

An estimate for copper ($\sigma = 6 \times 10^5$ ohm^{-1} cm^{-1}) at $v = 10^5$ cm/sec and $R = 1$ cm leads to $\alpha = 0.05$, i.e., to good conservation of the flux. This simple estimate shows that our assumption, namely, that in MC systems the magnetic flux is approximately conserved, is valid. We note incidentally that in the particular case when σ = const and $v \sim 1/R$, an analytic (self-similar) solution of the partial differential equation corresponding to a constant coefficient α, with $\Phi \sim R \sim t^{\alpha/2}$, has been obtained (M. P. Shumaev and the author, 1952).

Under real conditions the conductivity decreases strongly because of the Joule heating of the metal (all metals have a negative temperature coefficient of conductivity). A particularly important role is played by the "surface implosion" of the metal layer in which heat exceeding the sublimation energy is released. An expanding metal loses its metallic conductivity. If these processes result in production of a zone of reduced conductivity, comparable in magnitude with the radius of the cavity and having a conductivity such that the thickness of the skin layer in it is also comparable with the radius, then appreciable losses of the magnetic field occur. However, if the pressure of the magnetic field $H^2/8\pi$ and the density of the released Joule heat, which is comparable with it, are smaller than ρv^2, where ρ is the density of the shell, then the rate of the thermal expansion of the evaporated layer is smaller than $v = |dR/dt|$. The role of the surface implosion processes will then not be catastrophic even for an unfavorable conductivity of the plasma layer. The criterion

$$\rho v^2 > \frac{H^2}{8\pi} \tag{8}$$

comes into play also when we consider problems involving the stopping of the shell by the magnetic counterpressure, with allowance for the compressibility of the shell. At sufficiently large shell velocities, it is possible to attain arbitrarily high magnetic field energies. In the Soviet experiments the velocity of the cylindrical shells was 10–20 km/sec.

A question of practical importance is how to ensure very good cylindrical symmetry of compression of the metallic cylinder (i.e., high quality of the cylindrical charge) and of the detonation lenses, and precise simultaneity of initiation of the electric detonators. Slow residual deviations from symmetry increase upon compression, because of the dynamic instability of cylindrical-shell focusing. Therefore in practice the change in radius is limited to not more than a factor of 10. To obtain record fields it is therefore necessary to have very large initial fields. In Soviet experiments the MK-2 system was used to produce the initial fields (Fig. 6). Figure 7 shows an oscillogram obtained with a calibrated inductive pickup in an experiment in which a field of 25 million gauss was registered.

IV. APPLICATIONS OF MC

Measurements of the properties of substances in ultrastrong magnetic fields, obtained with the aid of MC, are made difficult by the short

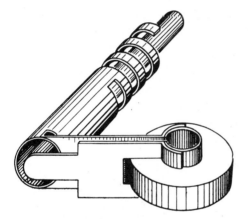

Figure 6 Use of an MK-2 generator to feed the core of an MK-1 generator.

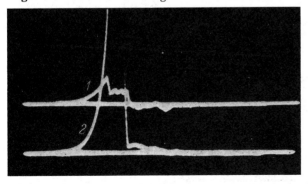

Figure 7 Oscillogram showing the magnetic field intensity. Beam 1: background (signal from shorted wires); beam 2: signal from measuring turn (integration with RC network).

duration of the process, by the presence of an air shock wave and of cumulative jets and particles in many cases, and by mechanical, thermal, and electric noise produced by the variable magnetic field itself. Special shielding of the measuring devices is necessary, and a pulse measurement technique must be used. In many cases it is advantageous to take the ultrastrong current outside the noise zone (Fig. 8, taken from Ref. 3, shows one of the devices of this type). Therefore the results actually obtained to date are not very numerous. Figure 9 shows one of the experimental setups for the measurement of the resistance of graphite in magnetic fields up to 1.5×10^6 Oe, obtained in a coaxial MC system, and Fig. 10 shows a characteristic oscillogram of the voltage on the sample (the increase in the voltage is due to the increase of the resistance of the sample and to the increase of the magnetic field intensity). Experiments were made aimed at observing the rotation of the

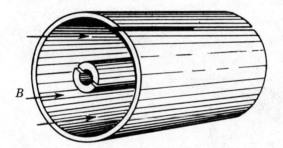

Figure 8   Magnetic field concentrator.

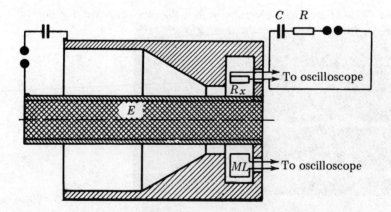

Figure 9   Diagram of experiment on the measurement of the resistance of graphite R_x sample, ML loop for measuring H; $RC \gg$ rise time of H.

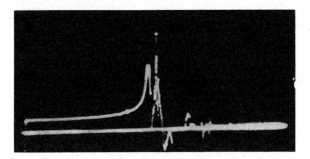

Figure 10 Oscillogram of voltage on a graphite sample.

light polarization plane in a magnetic field (Faraday effect) [2,3]. The very large change of light intensity observed in the experiments offers evidence that the plane of polarization was rotated by many thousands of degrees. American and Italian researchers have confirmed the linear character of the Faraday effect in glass in megagauss magnetic fields. The Americans obtained photographs of spectra of the Zeeman effect in fields up to two million gauss. The shift of the center of the Zeeman multiplet, resulting from compression of atoms by pressure of the magnetic field, was registered (the magnetic field is somewhat smaller inside the atom because of the atomic diamagnetism).

Soviet and American researchers have paid much attention to the use of MC systems for launching metallic bodies at escape velocities. This is of interest both for simulation of micrometeors and for experiments on physical processes at ultrahigh pressures, obtained when such bodies strike partitions. Figure 11 shows the diagram of a setup for

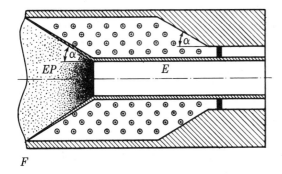

Figure 11 Diagram of setup for launching a ring.

launching an aluminum ring (weight ~2 g) which is accelerated by a pressure of a magnetic field in an annular gap to velocities exceeding 100 km/sec (this, to be sure, vaporizes the ring).

The most fundamental scientific application of MC generators, in our opinion, may be the supply of very high power to elementary-particle accelerators, and to measurement and recording apparatus. Let us consider a cyclic inductive accelerator constructed without an iron core (although this may not be the optimal design). The momentum of the accelerated proton is $p \sim HR$. The energy of the magnetic field is

$$W \sim R^3 H^2 \sim \frac{p^3}{H}$$

The proportionality coefficient can be determined by using as an example the already constructed coreless pulsed betatrons (see, for example, A. I. Pavlovskii et al. [10][†]). We obtain (the energy of the magnetic field is expressed in tons of TNT equivalent, 1 ton = 4×10^9 J)

$$W \text{ (tons)} = \frac{10^7 \text{ (G)}}{H} \left[\frac{P}{10^{10} \text{ (eV/c)}} \right]^3 \tag{9}$$

We see that in order to attain an energy of 1000 GeV = 10^{12} eV, which is the dream of modern high-energy physics, using $H = 10^7$ G at the center of the betatron (which certainly is not a limit), the required magnetic energy should be the equivalent of about one million tons of TNT. Obviously, the total energy should be several times larger, i.e., we are dealing with an underground explosion of a thermonuclear charge of "average" power. Such an explosion can be carried out without the spread of radioactive products at a depth somewhat greater than 1 km. The main expense is the construction, at such a depth, of a chamber with a volume larger than 10,000 m³, and the erection of several thousand tons of metal structures in such a chamber. We can count, however, on obtaining 10^{18} protons within a time of the order of 10^{-5} sec (the coefficient of utilization of the explosion energy is approximately 10^{-3}%). At such intensities we can obtain much scientific information in a single experiment. The repetition of the experiment 50–100 times will be comparable in cost with the creation of a continuously operating 1000-GeV accelerator (several billion rubles). Of course, it will be necessary to develop special recording apparatus, for example a system of photomultipliers registering the synchrotron radiation of the products of collision between the accelerated protons and the target, occurring in

---

[†] At an energy 5×10^5 J and a maximum field 1.2×10^5 Oe, the attained electron energy is 100 MeV.

special "measuring" magnetic fields in a large volume. From the radius of deflection in the magnetic field it is possible to determine in this case the momentum of the secondary particles, whose mass can be calculated from the spectrum and the intensity of the synchrotron radiation.

We must point also to another almost fantastic possibility. The use of large pulsed magnetic lenses (with magnetic field energy of hundreds of kilotons) makes it possible to focus a beam with an intensity $10^{18}/10^{-5} = 10^{23}$ protons/sec on an area of 1 mm^2. In this case it is possible to reliably register processes with colliding beams from two accelerators with a cross section of the order of 10^{-30} cm^2. To perform such experiments we must, of course, have automatic (feedback) systems to compensate for space charge and correct the magnetic field. Regardless of the extremely grandiose projects just described, it seems to us that MC generators can be useful in many fields of scientific research.

REFERENCES

1. C. M. Fowler, W. B. Garn, and R. S. Caird, *J. Appl. Phys. 31*:588 (1960).
2. R. S. Caird, W. B. Garn, D. B. Thomson, and C. M. Fowler, *J. Appl. Phys. 35*:781 (1964).
3. F. Herlach and H. Knoepfel, *Rev. Sci. Instrum. 36(8)*:1088 (1965).
4. A. D. Sakharov, R. Z. Lyudaev, E. N. Smirnov, Yu. N. Plyushchev, A. I. Pavlovskii, V. K. Chernyshev, E. A. Feoktistova, E. I. Zharinov, and Yu. A. Zysin, *DAN SSSR 165*:65 (1965); *Sov. Phys. Dokl. 10*:1045 (1966), trans. [Paper 3 in this volume].
5. Ya. P. Terletskii, *JETP 32*:387 (1957); *Sov. Phys. JETP 5*:301 (1957), trans.
6. E. I. Bichenkov, *PMTF 6*:3 (1964).
7. J. D. Lewin and P. F. Smith, *Rev. Sci. Instrum. 35*:541 (1964).
8. G. Lehner, J. G. Linhart, and J. P. Somon, *Nucl. Fusion, No. 4*:362 (1964).
9. F. Bitter, *Sci. Am. 213(1)*:65 (1965).
10. A. I. Pavlovskii, G. D. Kuleshov, G. V. Sklizkov, Yu. A. Zysin, and A. I. Gerasimov, *DAN SSSR 160*:68 (1965); *Sov. Phys. Dokl. 10*:30 (1965), trans.

Paper 5

Excitation Temperature in a Gas-Discharge Plasma

One of the most common methods of measuring the temperature of a gas discharge is the optical method based on measuring the relative intensities of appropriately chosen atomic lines or bands of molecular spectra. If there is thermodynamic equilibrium between the components of the plasma, the measured quantity is, naturally, the true temperature of the plasma. But if there is no thermodynamic equilibrium between the electrons on the one hand and the atoms and the molecules on the other, which is the case in a low-pressure plasma, it is necessary to consider what corresponds to the temperature measured by the optical methods, which is known as the "excitation temperature."

The concept of an excitation temperature can be introduced as follows. Suppose that we know the transition probabilities and intensities for a group of spectral lines. We can then find the concentrations N_i of the atoms (or molecules) in each excitation state and define for each group of lines the auxiliary quantity T_a (the excitation temperature) by setting

$$\frac{N_i}{N_j} = \frac{g_i}{g_j} e^{-(E_i - E_j)/kT_a}$$

from which it follows that

$$T_a = \frac{E_i - E_j}{k} \ln \frac{N_j g_i}{N_i g_j} \qquad (1)$$

Here, E_i and E_j are the excitation energies, and g_i and g_j the statistical weights of the upper and lower levels, respectively, of the corresponding transition.

Source: Температура возбуждения в плазме газового разряда, Известия АН СССР сер. физ. 12:372-375 (1948). Translated by J. B. Barbour.

Ladenburg and Kopfermann [1] have shown that if electron collisions are assumed to provide the main process of excitation of the atomic lines and one equates the number of electron collisions of the first kind to the number of collisions of the second kind (excitation and quenching), then the equation $T_a = T_e$, where T_e is the electron temperature, follows from the Klein–Rosseland relation.

It is well known that the Klein–Rosseland relation connects the cross sections S_{01} and S_{10} for collisions of the first and second kinds and is a consequence of quantum mechanical calculations. In the center-of-mass system of the two particles, we have

$$S_{01} p_0^2 g_0 = S_{10} p_1^2 g_1 \qquad (2)$$

where p is the momentum of the particle.

Suppose we have a nonisothermal mixture of particles of two species A and B; each species has its partial temperature T_A, T_B and mass m_A, m_B. In collisions with particles of type B, the particles of type A may be excited (E_1 is the excitation energy). In this case, one can find from (2) the ratio of the number of collisions of the first kind, Λ_{01}, to the number of collisions Λ_{10} of the second kind,

$$\frac{\Lambda_{01}}{\Lambda_{10}} = \frac{g_1 N_{A0}}{g_0 N_{A1}} e^{-E_1/k\tau} \qquad (3)$$

where

$$\tau = \frac{T_A m_B + T_B m_A}{m_A + m_B}$$

If the gas B is an electron gas, then $m_B \ll m_A$ and the "relative temperature" is $\tau \approx T_e$. If, further, $\Lambda_{01} = \Lambda_{10}$ (the assumption of Ladenburg and Kopfermann), comparison of (1) and (3) gives

$$T_a = T_e$$

However, it must be noted that under the actual conditions (ideal low-pressure gases) considered by Ladenburg and Kopfermann their assumption $\Lambda_{01} = \Lambda_{10}$ cannot be assumed to be satisfied (see, for example, Ref. 2).

At low electron concentrations, the deexcitation probability $\Lambda_f \sim 10^8$ sec^{-1} appreciably exceeds the probability of electron collisions of the second kind, which is equal to

$$\Lambda_{10} = \frac{1}{N_1} \int dN_1 \, dn_e \, S_{10} v_e \approx n_e S_{10} v_e$$

where n_e is the electron concentration, and v_e is the relative velocity of the electron and the atom.

The value of Λ_{10} becomes comparable with Λ_f only at fairly high electron concentrations n_e obtainable at several millimeters of mercury pressure (for example, if $n_e = 10^{14}$ cm^{-3}, $v_e = 10^8$ cm/sec, and $S_{10} = 10^{-14}$ cm^2, then $\Lambda_f = \Lambda_{10}$).

The region $\Lambda_{10} \ll \Lambda_f \approx \Lambda_{01}$ has been studied in detail by Fabrikant and his co-workers (see Refs. 3 and 4). Since the concentrations of the excited states in this region are proportional to n_e, the quantity n_e does not occur in (1) (it cancels), and the excitation temperature T_a is related to T_e by some functional dependence (but is not equal to it).

On the transition to pressures of the order of atmospheric pressure (where Λ_f can be ignored) we encounter the controversial problem of the relative importance of gas and electron collisions for the excitation temperature of the atomic lines. The estimate of Ornstein and Brinkman [5] corresponds to the conditions of an arc burning freely in the atmosphere. They find two numbers (the index g corresponds to the gas, and e to the electrons):

$$\Lambda_{10}^g = S_{10}^g n_g v_g$$
$$\Lambda_{10}^e = S_{10}^e n_e v_e$$

where

$$v_e = \sqrt{\frac{M_g}{M_e}} v_g = 3 \times 10^2 \text{ cm/sec}$$

They assume that S_{10}^g is approximately equal to S_{10}^e on the basis of experiments on the quenching of resonance fluorescence; for n_e they take the value $10^{-4} n_g$ (which from our point of view is somewhat stretched—the value of n_e may be 10–100 times greater). Ornstein and Brinkman then obtain $\Lambda_{10}^g \sim 30 \Lambda_{10}^e$ and conclude that molecular collisions play the dominant part. Quite apart from the dubious nature of the original numerical data (S_{10}^g, v_e) of this estimate, it must be emphasized that the ratio $\Lambda_{10}^g/\Lambda_{10}^e$ by itself says nothing. It is not only the number of collisions that is important but also the mean portion of energy that in each inelastic collision goes over from one form of energy (excitation energy) into another (kinetic energy). This mean portion Δ of the energy is very small in the case of resonance gas collisions ($\Delta_g \ll \Delta_e \approx E_1$). The value of E_1 (and Δ_e) is of order 1 eV, while Δ_g is smaller than 0.1 eV.

Instead of comparing the numbers Λ_{10}^g and Λ_{10}^e, it is more correct to compare expressions of the form

$$D_g = (\Delta_g)^2 \Lambda_{10}^g$$
$$D_e = (\Delta_e)^2 \Lambda_{10}^e \tag{4}$$

The second power of Δ in the expressions (4) can be justified as follows. If Δ is very small, the distribution of the molecules over the excitation levels satisfies a differential equation in which one of the terms has the form $D(d^2N/dE^2)$ [where D is defined in (4)].

In the case of large $\Delta \sim E$, we cannot go over from the algebraic equation for N_i to a differential equation, but it is natural to use D as before to characterize the intensity of the process of mutual transformation of the energy forms. The use of (4) leads to the conclusion that the mechanism of electron collisions plays the predominant part under real arc conditions (Δ_e is tens of times greater than Δ_g, and Λ_{10}^g cannot be hundreds of times greater than Λ_{10}^e).

Thus, a measurement of the excitation temperature for atomic lines gives us information about the electron temperature. It is necessary to know only if one can ignore under the given conditions deexcitation or, on the other hand, collisions of the second kind, or, finally, whether we have an intermediate case. In the first case, the excitation temperature is directly the electron temperature of the plasma. In the remaining cases, the excitation temperature does not have such a direct meaning.

We now consider the physical meaning of T_a for electron-rotation bands. The probabilities w of electron-rotation transitions $k_0 \to k_1$ depend in a simple manner on the rotational quantum number k_0 [$w_R \sim k_0/g_0$ for $k_1 = k_0 - 1$, and $w_P \sim (k_0 + 1)/g_0$ for $k_1 = k_0 + 1$].

In the usual method of finding T_a one plots $\ln(I/wg_0)$ (I is the intensity of the spectral line) as a function of $k_0(k_0 + 1)$. From the slope of the resulting straight line one obtains T_a. It is remarkable that experimentally one obtains a well-defined straight line even when the gas pressure is so low that the deexcitation time $1/\Lambda_f \sim 10^{-8}$ sec is appreciably shorter than the mean time $1/S_g n_g v_g$ between two collisions of molecules (see, for example, the experiments of Ref. 6: the nitrogen pressure was 0.1 mmHg; the excitation was by means of electrons with an energy of about 150 eV; and T_a was found to be about 500 K).

It is obvious from these experiments that excitation by electron collision does not significantly affect the Boltzmann distribution with respect to k (the distribution of the molecules with respect to the rotational levels in the upper electron state copies this distribution in the lower state). It appears to us that the reason for this is as follows. The maximum angular momentum $\delta = k_0 - k_{00}$ that an electron can transfer to a molecule in a collision is (in units of $\hbar$) $2Rp_e = 2R/\lambdabar_e$, where $2R$

is the diameter of the molecule, and $\lambda_e = 1/p_e$ is the de Broglie wavelength for the electron. This estimate follows from classical arguments and also from the expressions of the Born approximation.

Further, we made the following approximate calculations. We approximated the dependence of the cross section S_{01} on k_0 and δ by a function convenient for calculation; we took $\delta_{\max}$ equal to 5, which is reasonable for the conditions of the experiment in Ref. 6. The distribution with respect to k_{00} was assumed to be a Boltzmann distribution with $T_a = 500$ K. It was found that for $k_0 \geq 5$ the distribution with respect to k_0 (after electron collision) is also well reproduced by the Boltzmann expression with $T_a = 500$ K. Assuming that the nature of the copying property of the electron collisions is as elucidated, we can now consider the factors that determine the distribution with respect to k_{00}. It is again necessary to find D_g and D_e in accordance with (4). Massey [7] found that the cross section for inelastic collision of an electron with a diatomic molecule can be of the order of the gas-kinetic cross section if the molecule has a dipole moment. However, k_{00} is then changed by only 1, i.e., Δ_e is much smaller than $\Delta_g \sim kT$. This has the consequence that in a wide range of conditions the inequality $D_g \gg D_e$ is satisfied, i.e., equality of T_a and the gas temperature is ensured. Thus, we obtain the possibility of measuring the gas temperature by means of the distribution of the intensities in an electron-rotation band.

I am very grateful to S. L. Mandel'shtam for suggesting the subject and for numerous discussions.

REFERENCES

1. H. Kopfermann and R. Ladenburg, *Naturwissenschaften* 19:512 (1931).
2. K. Panevkin, *ZhETF* 9:1007 (1939).
3. V. A. Fabrikant, *Dokl. Akad. Nauk SSSR* 15:451 (1937).
4. V. A. Fabrikant, *Uspekhi. Fiz. Nauk* 27:1 (1937).
5. S. Ornstein and H. Brinkman, *Physica* 1:797 (1934).
6. A. E. Lindh, *Z. f. Phys.* 67:67 (1931).
7. H. S. W. Massey, *Proc. Cambridge Philos. Soc.* 28:99 (1932).

Commentary: Harold P. Furth*

Controlled Fusion Research

The fundamental scientific paper of the Soviet controlled fusion research effort was written in 1950 by A. D. Sakharov and I. E. Tamm. This paper became available to the international scientific community in 1958, through the general declassification of controlled fusion work.

The Sakharov–Tamm paper is addressed mainly to the problem of producing net power from a hot deuterium plasma confined in a magnetic bottle. The second part of this three-part paper was written by Sakharov; it offers a number of basic physical insights that have retained their validity to this day, and describes a particular type of toroidal bottle that later rose to fame under the name *tokamak*. Sakharov also made brilliant and fundamental suggestions in two other research areas related to controlled fusion: the generation of ultrahigh magnetic fields and the catalysis of fusion reactions by mu mesons.

TOROIDAL MAGNETIC BOTTLES

The possibility of nonexplosive fusion-power generation arises from the circumstance that high-temperature matter decomposes into a plasma of charged particles, electrons and ions, which can be restricted to localized orbits by a strong magnetic field. The basic charged-particle orbit is helical: particles gyrate around magnetic field lines and move freely along them. The choice of a toroidal shape is therefore natural for a magnetic bottle: it prevents the escape of particles along magnetic field lines. Sakharov's paper points out, however, that a *purely* toroidal magnetic field is an unsatisfactory solution, since the particle orbits exhibit a slow sideways "drift" in the direction of the axis of symmetry.

The remedy proposed by Sakharov was to add a weak poloidal

*Plasma Physics Laboratory, Princeton University, Princeton, New Jersey magnetic field component, thus creating a set of nested toroidal magnetic surfaces (somewhat like a doughnut-shaped onion), which contain helical magnetic field lines. The particle orbits then become localized to the vicinity of the magnetic surfaces, and the hot plasma can be isolated effectively from the cold walls of the fusion reactor.

How can the essential poloidal-field component be created? Sakharov proposed two solutions:

1. The toroidal plasma contains and completely surrounds a current-carrying ring; the toroidal ring current generates poloidal magnetic field and is supported by this field against gravity. Sakharov envisaged the possibility of maintaining the ring current in the presence of resistive dissipation by transmitting power to it through radio waves, but foresaw great technical difficulties. During the 1960s experimental devices of the internal-ring type were actually built in the U.S., under the names *levitron* and *spherator*. Thanks to the advent of modern superconductor technology, the current of levitated rings weighing hundreds of pounds was maintainable for periods of many hours. Sakharov's floating-ring approach is still considered to be feasible and even attractive from the point of view of plasma confinement—but the technical problems of a fusion reactor following this concept are also still considered to be unmanageable in the near term.

2. Sakharov's alternative suggestion was to induce a toroidally directed current within the plasma itself—which is an excellent electrical conductor. This approach later came to be the main experimental thrust of the Soviet fusion research program. Towards the end of the 1960s, under the leadership of the late Academician L. A. Artsimovich, the Soviet tokamak effort was able to reach temperatures of about 10 million °K and plasma confinement times of a hundredth of a second—a success far beyond all prior experimental achievements of world fusion research.

The declassification of controlled fusion research, in 1958, revealed a remarkable degree of resemblance among the secret fusion reactor concepts of the U.S.S.R., the U.K., and the U.S. All three programs had developed tokamak-like toroidal magnetic bottles, but there were significant divergences of insight and emphasis.

The beginnings of fusion-related British plasma research were published in the late 1940s by S. W. Cousins and A. A. Ware, who worked with a toroidal magnetic bottle composed of *purely poloidal field*. This "dynamic pinch" was, in principle, an excellent confiner of plasma particles, but suffered in practice from gross magnetohydrodynamic instabilities. Subsequently, a toroidal magnetic field component was added in pursuit of the "stabilized pinch" concept, but the very high degree of stability that characterizes the tokamak, with its predominantly toroidal magnetic field, still remains to be achieved by pinch research.

The early U.S. fusion program also pursued the pinch effect, but developed a number of other approaches as well. In 1951, the Princeton astrophysicist L. Spitzer, Jr., had set himself the same problem as

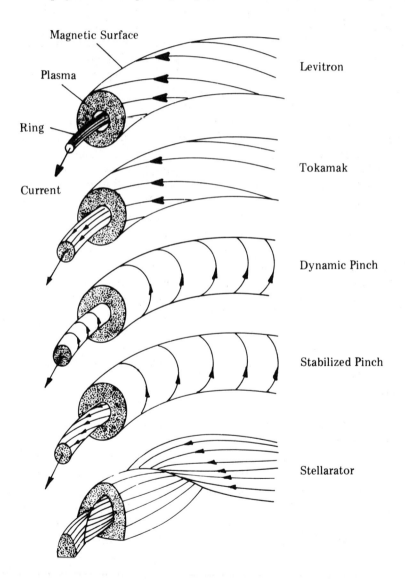

Sakharov—the improvement of the pure-toroidal-field bottle through introduction of magnetic surfaces—and had also come up with two solutions: (1) the "stellarator," which superimposes an *average poloidal field component* on a strong toroidal field by means of currents in helical multipole windings external to the plasma; and (2) the tokamak, which Spitzer considered to be a less desirable alternative. The Princeton experiments compared stellarators, tokamaks, and tokamak-stellarator hybrids, but did not achieve parameters as favorable as those of the Soviet tokamak experiments.

The status of Sakharov and Tamm as the principal forefathers of modern tokamak research derives partly from the outstanding experimental success of the Soviet program, and partly from their highly perceptive and relevant insights into tokamak physics. The role of ion thermal transport as a dominant energy loss mechanism was identified in the 1951 paper, and was analyzed by Tamm in a manner that came very close to modern neoclassical transport theory. (Sakharov anticipated another complex phenomenon that is important to tokamak experiments, the enhanced penetration of neutral atoms into the hot plasma interior, through the mechanism of multiple charge-exchange events.) The concern that Sakharov expressed about the destabilizing potential of even a weak current flowing within the plasma was subsequently formalized into the Kruskal–Shafranov limit of magnetohydrodynamic stability theory, which imposes a critical constraint on the relative strengths of the poloidal and toroidal field components.

At the present time, the tokamak approach has become the main line of international fusion research. Four very large tokamak experiments, the Joint European Torus (JET) in England, the Tokamak Fusion Test Reactor (TFTR) in the U.S., the JT-60 in Japan, and the T-15 in the U.S.S.R., are planning to obtain "break-even plasma conditions" during the mid-1980s. Looking further into the future, the tokamak has a potential drawback: steady-state reactor operation is technically very desirable, but may prove to be unattainable, because the duration of the transformer-induced toroidal current pulse is limited.

The basic drawback of the simple tokamak, which was noted by Sakharov in his 1951 paper, will need to be overcome through the introduction of advanced design features. One option is the use of radio waves to drive a steady-state tokamak plasma current (in loose analogy with Sakharov's suggestion for maintenance of a levitated-ring current). Recent experimental successes of the stellarator approach have also rekindled interest in fusing the original insights of Sakharov and Spitzer in the form of a tokamak-stellarator hybrid reactor.

ULTRAHIGH MAGNETIC FIELDS

During the time of Sakharov's work on the toroidal magnetic bottle, he inspired another important area of research that has obvious application to controlled fusion. In 1951 Sakharov suggested that extremely high magnetic fields could be generated by using an explosive charge to compress a metal tube containing axial magnetic flux. This flux could be generated slowly by means of an external solenoid, and would become trapped and compressed during the implosion, ultimately producing a very strong magnetic field in a small volume.

This scheme was tried experimentally in 1952 by Sakharov's colleagues R. Z. Lyudaev, E. A. Feoktistova, G. A. Tsyrkov and A. A. Chvileva, and succeeded in producing fields above a megagauss—three times the record magnetic field strength attained by P. Kapitza at Cambridge in 1927. During the late 1950s, C. M. Fowler, W. B. Garn, and R. S. Caird, at the Los Alamos Scientific Laboratory, reached the 10-megagauss level by the same method and reported their work in the open literature. In 1965, the Sakharov group began to publish, and described the attainment of maximum field strengths above 25 megagauss. The achievement of such extremely strong fields evidently represents a step towards physical conditions normally associated only with astrophysical phenomena: the Maxwell stress exerted by a 25-megagauss field corresponds to a pressure of 25 million atmospheres!

In regard to potential applications of the explosive high-field generator, Sakharov suggests "relatively small single-shot accelerators to produce 100–1000 GeV, as well as plasma studies, acceleration of dense bodies to speeds around a thousand kilometers per second, etc." The possible application to fusion "plasma studies" in the form of compression to ignition is obvious, and has been studied extensively as the basis for a simple, pulsed fusion reactor. The preferred modern version of this approach makes use of magnetic field pressure or gas pressure to implode a liquid-metal tube containing a fusion plasma and its magnetic bottle.

MU-MESON FUSION

Sakharov's uncommonly wide range of competence as a physicist is illustrated by yet a third contribution to controlled fusion research, dating from approximately the same period as the others. Sakharov and his colleague Ya. B. Zel'dovich were the first to recognize and pursue the amazing potential of the mu meson for catalyzing fusion reactions in *cold* deuterium.

At the time of discovery of the process of pi meson decay into what seemed to be another meson, in 1947, F. C. Frank of the University of Bristol examined the plausibility of various alternate explanations. He proposed the idea that a slowed-down pi meson might form a molecule with one deuterium nucleus and one ordinary hydrogen nucleus in the target emulsion, and calculated that these tightly bound nuclei would then readily fuse, while the pi meson would be ejected with appreciable energy. He also showed, however, that the small natural abundance of the deuterium isotope made this process highly unlikely. The authenticity of the pi-meson decay process was sustained, and the mu meson was recognized as a new elementary particle of appreciable lifetime (2.2 μsec).

While Frank had proposed pi catalysis as a possible source of scientific confusion, Sakharov was apparently the first to propose mu catalysis as a possible technique for releasing fusion power in a cold medium. (The published literature contains only a reference to "A. D. Sakharov, Report of the Physics Institute, Academy of Sciences, 1948.")

In 1956, the phenomenon of mu-meson fusion was discovered experimentally in a hydrogen bubble chamber at the Berkeley Bevatron by L. Alvarez and his co-workers—who mistook it at first sight for a possible new mode of mesonic decay, but soon deduced the true nature of the process. Theoretical work published by Zel'dovich in 1954 was then found, and the possibility of a cold mu-catalyzed deuterium-tritium fusion reactor became a tantalizing subject of analysis.

The obstacles to this type of reactor are twofold: (1) in liquid deuterium-tritium, a mu meson could catalyze only about a hundred fusion reactions during its lifetime; and (2) unfortunately, the mu meson also has a probability of 1% of getting stuck on one of the helium decay products of the fusion reaction, thus being effectively retired from further catalytic action. The first problem could be relieved by operating at superhigh density, but the second seems to be inherent. The energy balance of the mu-catalyzed reactor is so remarkably close to being economically attractive that the intellectual quest for some practical solution has not been abandoned to this day—and might even succeed in the long run.

Commentary: Thomas H. Stix*

Plasma Physics

A set of four volumes, edited by Academician M. A. Leontovich and presented to the international scientific community in 1958, collects and summarizes research in plasma physics and controlled thermonuclear reactions done in the Soviet Union during the 1950s. The editor points out that the grouping into volumes corresponds roughly to the chronological order of the work which they report. With this in mind, it is especially significant that the first presentation in the first of these four volumes is a three-part paper by Tamm and Sakharov. Parts I and III, by Academician I. E. Tamm, are principally devoted to basic principles of the physics of high-temperature magnetized plasmas, including orbit theory, kinetic theory, collisions, heat transport, and viscosity. Part II of this ground-breaking Soviet work was written by Academician Sakharov and appears, in translation, in this collection. Its title is "Theory of Magnetic Thermonuclear Reactor" and it introduces the idea of power balance (later quantified in the well-known Lawson criterion) between thermonuclear energy production and bremsstrahlung loss. But even more interesting to today's plasma physicists are Sections III and IV of Part II in which the concepts of toroidal magnetic confinement of a fusion plasma are laid out with impressive insight. Academician L. A. Artsimovich describes this work as follows [1].

> In the Soviet Union, the investigation of the problem of magnetic traps was given impetus by the work of Sakharov and Tamm who, in 1950, proposed the first model of a thermonuclear reactor based on magnetic confinement. They proposed to ionize and heat a rarefied deuterium gas confined within a toroidal chamber by a strong longitudinal magnetic field generated by coils wound on the outer surface of the chamber. However, such a system has an essential defect as a

*Plasma Physics Laboratory, Princeton University, Princeton, New Jersey

magnetic trap: in the nonuniform toroidal field, the particles have a drift motion perpendicular to the magnetic lines of force.... This motion will cause each particle to impinge on the wall of the chamber. This defect was first pointed out by Sakharov and Tamm themselves, who then proposed to eliminate the drift by using either the self-magnetic field of a longitudinal plasma current, or the field of a coaxial current-carrying ring suspended freely inside the chamber.

The latter method describes the class of configurations later built and tested under the names *levitron* and *spherator*. The former method is just exactly the tokamak, considered today the front-runner among the means proposed to achieve a practical fusion reactor.

Among the ways to achieve controlled fusion, we must certainly mention one approach to the problem which caused great excitement in the 1950s and which "almost" works. The suggestion, first made by Sakharov [2], is that mu mesons could be used to catalyze the d-d fusion reaction. As the mu meson replaces the electron in the molecular ion $(dd)^+$, its large mass shortens the interionic distance in the molecule to about 5×10^{-11} cm. Reduction of the Gamow barrier then reduces the d-d fusion lifetime down to approximately 10^{-11} sec, which can be compared to the mu-meson decay lifetime of 2.2×10^{-6} sec. Unfortunately, the mu-meson catalyst still disappears by decay—owing to the rather long time for mesic atom formation and, more important, because of the finite probability of mesic attachment to the fusion-product helium nuclei—at a rate too fast for a favorable power balance. Nevertheless, Sakharov's suggestion has recently been revived by Petrov [3] in the context of a fusion-fission hybrid scheme where neutrons from mu-meson d-t catalysis would be used to induce fission of ordinary uranium, U^{238}.

Sakharov's interest in the practical details of experimental work is illustrated in a totally different way in a short paper, "Excitation Temperature in a Gas-Discharge Plasma" [Paper 5]. Attention is paid here, we think for the first time, to the interpretation of spectral-line intensity ratios when the plasma electrons have not come to thermodynamic equilibrium with the atoms and molecules of the gas. The problem which Sakharov attacks is still of current interest in the study of modest-temperature low-density plasmas such as occur at the plasma periphery and near the vacuum chamber wall in controlled fusion experiments.

Finally, in this brief review, we come to two papers on the imaginative use of the energy of an explosion to produce ultrastrong (1–20 MG) magnetic fields. That Sakharov was the originator of this concept is made clear in the second sentence of the 1966 multiauthor paper, reprinted here, entitled "Magnetic Cumulation" [Paper 3]. Applications which Sakharov suggests for this capability include the measurement of physical properties (e.g., resistivity, Faraday effect, Zeeman effect), acceleration of macroscopic metallic objects, a super betatron with a 10-MG central field for particle acceleration to 10^{12} eV (a confined nuclear explosion is suggested to power this experiment), and superpowerful magnetic lenses to focus colliding beams to an area of 1 mm^2 for the study of processes with cross sections of the order of 10^{-30} cm^2. Again in the instance of high-magnetic-field production, it can be pointed out that a field first opened up by Sakharov still attracts great interest: Session BC, in New York City on January 26, 1981, of the 1981 Annual Meeting of the American Physical Society consists of four invited papers on the topic "Megagauss Physics." With respect to Sakharov's initiative, it is of special interest to note that paper BC 3 by R. S. Caird of the Los Alamos Scientific Laboratory, is entitled "Applications of Explosively Generated Megagauss Fields."

REFERENCES

1. L. A. Artsimovich, *Controlled Thermonuclear Reactions* (A. C. Kolb and R. S. Pease, eds; P. Kelly and A. Peiperl, trans.), Oliver and Boyd, Ltd., Edinburgh and London, 1964, page 237.
2. A. D. Sakharov, *Report of the Physics Institute*, Academy of Sciences, USSR (1948). This original work is cited, for instance, in the interesting review article on mesonic catalysis by Ya. B. Zel'dovich and S. S. Gershtein, *Uspekhi Fiz. Nauk* 71:581 (1960); *Sov. Phys. Uspekhi* 3:593 (1961), trans.
3. Yu. V. Petrov, *Nature* 285:466 (1980).

Part 2
COSMOLOGY

Commentary: A. D. Sakharov

Cosmological Investigations

1. "The initial stage of an expanding universe and the appearance of a nonuniform distribution of matter" [Paper 6]. The gravitational instability of the "cold" model of the universe is considered. The results of previous studies of E. M. Lifshits and other authors for the classical instability are repeated and generalized with allowance for a pressure gradient. One of the results is a general formula for perturbations in which one can ignore the effects of the pressure gradient (wavelength small compared with the cosmological radius of curvature). Taking

$$\frac{\Delta \rho}{\rho} = \frac{-3a_1}{a_0}$$

(a_0 is the unperturbed "scale" and a_1 are infinitesimally small variations of the "scale"), we have for a_1,

$$\frac{\ddot{a}}{a_1} = \frac{\dddot{a}}{\dot{a}_0}$$

(the right-hand side is found from the condition that a_1 is the solution corresponding to a displacement of a_0 in time).

The quantum problem is investigated by means of a self-similar solution for a harmonic oscillator with variable parameters (generalized mass m and generalized elasticity coefficient R). The wave function $\psi(z,t)$, where $z \sim \Delta\rho/\rho$ is sought in the form $\psi = v(t)\exp(-\mu(t)z^2)$ and for $\mu(t)$ the following ordinary differential equation is solved:

$$\frac{d\mu}{dt} = i\left(\frac{2\mu^2}{m} - \frac{r}{2}\right)$$

The main difficulty was in reducing the equations for the perturbations to Hamiltonian form. For this, an infinitesimally small transformation of the coordinate system was used: $t \to t + \Delta t$, $(d/dt)(\Delta t) = -z(dp/d\varepsilon)$, where p is the pressure and ε the energy density.

At present, the "hot" model of the universe is the most popular one. However, in my opinion, a number of results of the paper under discussion are also of interest in this case. The main point to note, however, is the following. It is quite possible that the earliest stages in the development of the universe should still be described by the "cold" model (see, for example, 6 below and Paper 12).

2. "Violation of CP invariance, C asymmetry, and baryon asymmetry of the universe" [Paper 7]. This paper makes a suggestion concerning the origin of the observed baryon asymmetry $A = N_B/N_\gamma \sim 10^{-9}$ of the universe from an initial charge-symmetric state as a result of nonequilibrium processes in the early stages in the expansion of a "hot" universe with violation of CP invariance and nonconservation of baryon charge. Violation of CP invariance had been discovered experimentally not long before the paper was written. Nonconservation of baryon charge was postulated in the paper and a definite mechanism was proposed to ensure conservation of the "combined charge" $3B + L$ (in the paper, L_μ).

It was pointed out that, because of CPT symmetry, baryon asymmetry cannot arise under stationary conditions, so that departures from equilibrium due to the cosmological expansion are important. The combined law of conservation for the baryons (with allowed quark-lepton transformations) leads to much longer proton lifetimes at the same mass of the quark-lepton boson than is predicted in the SU_5, O_{10}, etc., grand unification schemes. Therefore, if it is found that the proton decays with a lifetime of the order of 10^{30} years, my hypothesis of a combined conservation law must be abandoned.

The paper also puts forward the idea of cosmological CPT symmetry with respect to the Friedmann singularity. Such a symmetry includes a change in the direction of the flow of entropy time. The "reversal of the arrow of time" eliminates the paradox of irreversibility known since the end of the nineteenth century. For $t < 0$, the time derivatives occur in the statistical equations with the opposite sign, while for $t > 0$ they occur with the usual sign, i.e., in the cosmological theory the symmetry between the two directions of time inherent in the equations of motion is also recovered for nonequilibrium processes (including life processes).

From cosmological CPT symmetry it follows that the currents at the moment of a singularity disappear. Hence, the universe should be neutral at all conserved charges. This consideration was precisely the starting point of my work.

Nonconservation of the baryon charge was proposed in 1964 by S. Weinberg, who pointed also to its possible connection with baryon asymmetry. In his paper however, there was no mention of the role of nonstationarity effects.

3. "Antiquarks in the universe" [Paper 9]. This appeared in a collection of papers in a volume to mark the 60th birthday of N. N. Bogolyubov in 1969. The paper is based on the hypothesis of quarks with integral charges and from my point of view is no longer of interest. An attempt is made to return to a baryon conservation law, and it is suggested that the antibaryon charge is concentrated in the from of not readily observable neutral stable quarks with a mass of several GeV.

4. "A multisheet cosmological model" [Paper 10]. This appeared as a preprint of the Institute of Applied Mathematics, Moscow, in February 1970. Expanding and contracting spaces are "joined together" by gravitational collapse. It is assumed that a black hole in an expanding space passes through a singular worldline into a white hole in a contracting space. In the spatially flat four-dimensional world, the entire mass of the matter in the universe undergoes collapse in the limit as $t \to \infty$. An infinite sequence of such sheets is cosidered, this giving rise to the name of the model. A shortcoming of the model is the high degree of inhomogeneity that arises in a contracting space. It is probable that this cannot be readily reconciled with the observed degree of homogeneity of the universe. In Papers 5 and 6 the same name is applied to a pulsating model with nonzero cosmological constant.

5. "The baryonic asymmetry of the universe" [Paper 11]. In 1978, Yoshimura again advanced the idea that baryon asymmetry could arise through violation of CP invariance and nonconservation of the baryon charge in a unified theory of the strong, weak, and electromagnetic interaction (schemes of the SU_5 type, in which the proton is unstable against the reaction $p \to e^+ + \pi^0$). In my paper, I developed the ideas which I had put forward in 1967. I assume the existence of a combined conservation law for $3B + L$. According to an estimate, A does not depend explicitly on the mass of the quark-lepton boson, but one requires the presence of quarks or Higgs bosons with masses of the same order. Estimates are made for a theoretical model that differs strongly in a number of respects from an SU_5 scheme (there is a combined conservation law for $3B + L$, and three types of quark-lepton bosons are postulated). Some aspects of the multisheet cosmological model (modified as compared with 6 by having a nonvanishing cosmological con-

stant) are considered. If the curvature and invariant charge are zero, the statistical characteristics (for example, A) are repeated from cycle to cycle and can be determined from this requirement.

6. "Cosmological models of the universe with reversal of time's arrow," [Paper 12]. The hypothesis of cosmological CPT symmetry is again advanced, and attention is drawn to a possible departure from it when a cosmological invariant charge is present. The main aim of the paper is to combine the hypothesis of the multisheet model and the hypothesis of reversal of the arrow of time. In the model, I propose a negative spatial curvature of the universe and a finite cosmological constant whose sign corresponds to a negative vacuum energy density ε of flat space. The hyperbolic radius a for the cycles of contraction and expansion on either side of the time of reversal of the arrow of time changes in accordance with the law

$$a = a_0 \sin a_0^{-1} t$$

$$a_0 = \left(\frac{8\pi}{3} G |\varepsilon|\right)^{-1/2}$$

As entropy and the mass of the matter proportional to it accumulate, there is a gradual transition to the asymptotic regime

$$a = a_n \left(\sin \frac{3 a_0^{-1} t}{2}\right)^{2/3}$$

The maximum hyperbolic radius of cycle n, $a_n > a_0$, is determined by the condition that the matter density satisfy $\rho = m_p A n_\gamma = |\varepsilon|$ (m_p is the proton mass and A is the baryon asymmetry). With increasing number of photons, a_n increases in accordance with a power law, and the curvature is $1/a_n^2 \to 0$ as $|n| \to \infty$. The pulsation period tends to $2\pi a_0/3$. The asymptotic regime corresponds to the multisheet cosmological model described in 5 without reversal of the arrow of time.

The small dimensionless quantity $\delta^2/a^2 \sim 10^{-58}$, where $\delta^{-3} = n_\gamma$, which characterizes the mean spatial curvature of the universe, is explained as the result of the accumulation of entropy during many successive cycles of expansion and contraction. The initial entropy at the point of reversal of the arrow of time can be taken to be zero, i.e., the model can be cold.

7. "Maximal temperature of thermal radiation" [Paper 13]. The relevance of this paper is diminished by the fact that under the limit densities there are no stationary states, and one can speak about temperature only in the sense of the derivatives $(\partial E/\partial S)|_V$.

Paper 6

The Initial Stage of an Expanding Universe and the Appearance of a Nonuniform Distribution of Matter

A hypothesis of a creation of astronomical bodies as a result of gravitational instability of the expanding universe is investigated. It is assumed that the initial inhomogeneities arise as a result of quantum fluctuations of cold baryon-lepton matter at densities of the order of 10^{98} baryons/cm^3. It is suggested that at such densities gravitational effects are of decisive importance in the equation of state, and the dependence of the energy density ε on the baryon density n can qualitatively be described by graphs a and b of Fig. 1. The quantity ε vanishes at a certain density $n = n_0$. A theoretical estimate (containing some vague points) yields initial inhomogeneities in the distribution of matter which can explain the origin of clusters of $10^{62}-10^{63}$ baryons (10^5-10^6 $M_\odot$). The calculated mass is smaller than that of the galaxies by a factor of 10^5-10^6; it is in fact closer to the masses of globular clusters. The hypothesis is proposed that galaxies are produced as a result of an increase of nonuniformities in the motion and distribution of the gas formed during gravitational collapses of the primordial stellar clusters. According to this hypothesis the plane component of the galaxy is produced from the gas, whereas the spherical component consists mainly of clusters of primordial stars captured by the gravitational field of the rotating gas cloud.

UNITS AND NOTATION

In the gravitational system of units we put $\hbar = c = G = 1$.

The unit of length is $c^{-1/2} \hbar^{1/2} G^{1/2} = 1.61 \times 10^{-33}$ cm.

The unit of time is $c^{-3/2} \hbar^{1/2} G^{1/2} = 5.35 \times 10^{-44}$ sec $= 1.7 \times 10^{-51}$ year.

The unit of mass is $c^{1/2} \hbar^{1/2} G^{-1/2} = 2.18 \times 10^{-5}$ g $= 10^{-38}$ $M_\odot$.

The baryon density is n; $n = 1$ at a density 0.24×10^{99} baryon/cm^3.

Source: Начальная стадия расширения Вселенной и возникновение неоднородности распределения вещества, ЖЭТФ 49:345-358 (1965). Reprinted from *Sov. Phys. JETP* 22:241-249 (1966), translated by J. G. Adashko, with permission from American Institute of Physics.

The average distance between baryons is $a = n^{-1/3}$. The relative perturbation of the density is expanded in a Fourier series of the orthogonal functions $\Phi_\varkappa$, defined (to simplify the normalization) in a cube of the volume period $V = V_0 a^3$ (V_0 does not depend on the time):

$$\frac{\Delta n}{n} = \sum_\varkappa \Phi_\varkappa(\xi) z_\varkappa(t) \tag{1}$$

Here $\xi = x/a$ is the comoving dimensionless spatial coordinate (3-vector), and $\varkappa$ is a dimensionless wave vector (independent of the time). The function Φ satisfies the normalization condition and the equation

$$\int_{V_0} |\Phi|^2 d\xi = 1$$

$$\Delta \Phi + \frac{\varkappa^2}{a^2} \Phi = 0$$

For example, $\Phi = v_0^{-1/2} \exp(i\varkappa\xi)$. The quantity $z_\varkappa$ defined in accordance with (1) is called the *amplitude of the inhomogeneity*.

ε is the energy density in the rest frame of the material.

$p = n \, d\varepsilon/dn - \varepsilon$ is the pressure.

A dot over a letter denotes a time derivative.

C is the limit of $\dot{a}^2$ as $t \to \infty$; the symbol $\sim$ stands for equal in order of magnitude; the symbol $\backsim$ denotes proportionality.

I. INTRODUCTION

The cosmological theory of the expanding universe is at present generally accepted. This theory is based on a nonstationary solution obtained by A. A. Friedmann for Einstein's equations of general relativity, and explains, in particular, the phenomenon of the "red shift" (see, for example, the series of articles dedicated to the memory of A. A. Friedmann [1]).

Recently Ya. B. Zel'dovich (see, for example, Ref. 1) presented convincing arguments in favor of the assumption that matter was cold in the initial, dense state and indicated that under certain assumptions concerning the initial relations between the baryon and lepton densities it is possible to explain, within the framework of these ideas, the predominant abundance of hydrogen in the universe and the low temperature of intergalactic space. It can be assumed that during the early stage of the expansion, the matter in the universe was practically homogeneous, and the "primordial" astronomical objects were the result of gravitational instability. Although many astronomers and astrophysicists object to this point of view, it is essential to investigate it. For the development of such a hypothesis, great significance is attached to a study of the laws whereby small inhomogeneities of density build up, and to the determination of the statistical characteristics of the initial inhomogeneities. The first problem was solved within the framework of the theory of the expanding universe by E. M. Lifshits and again considered by Ya. B. Zel'dovich (see Refs. 1 and 2), while the solution of the second problem calls for an examination of the initial stage of expansion of the universe, and therefore contains many hypothetical premises. However, by combining theoretical considerations and astronomical observations we can hope to obtain the necessary initial data for the solution of cosmogonical problems and information concerning the physical laws obtaining at extremely high densities of matter.

The fundamental equation of the theory of the expanding universe can be taken to be one of Einstein's equations:

$$R_0^0 = 8\pi(T_0^0 - \tfrac{1}{2}T)$$

Substituting for the case of a homogeneous world $R_0^0 = -3\ddot{a}/a$ and the components of the energy-momentum tensor in the form $T_0^0 = \varepsilon$, $T = \varepsilon - 3p$, we get

$$\ddot{a} = -\frac{4\pi}{3}(\varepsilon + 3p)a \tag{2}$$

Equation (2) has the integral

$$\dot{a}^2 = \frac{8\pi}{3}\varepsilon a^2 + C \tag{3}$$

where the integration constant C determines, as is well known, the sign and the magnitude of the spatial curvature. If we take Oort's estimate for the average density of matter at the present time $\bar{\rho} = 3 \times 10^{-31}$ g/cm^3, then $\dot{a}^2 > 8\pi\varepsilon a^2/3$ and the constant $C > 0$, although it is very small: $C \approx \dot{a}^2 = 1.6 \times 10^{-52}$. This is a case of Lobachevskiĭ geometry. However, estimates of $\bar{\rho}$ are not very accurate and it is not excluded that $C = 0$ and $\bar{\rho} = 1.2 \times 10^{-29}$ g/cm^3, and that at the present time $\dot{a}^2 = 2 \times 10^{-53}$ ($\dot{a}^2 \to 0$ as $t \to \infty$), and even $C < 0$ is not excluded.

It is assumed in this paper that the equation of state of matter at $n \ll 1$ coincides qualitatively with that for a degenerate Fermi gas, and that the effect of interaction and transformation of fermions does not play a very important role. Accordingly, $\varepsilon \sim n^{4/3}$ for $n^{1/3} > M$ ($M = 0.76 \times 10^{-19}$ is the baryon mass) and $\varepsilon = Mn$ when $n^{1/3} < M$.

We distinguish arbitrarily between four (or three) stages of the expansion of the universe:

stage 1	$\varepsilon \lesssim 1$	$a \lesssim 1$	$1 \gtrsim t$
stage 2	$\varepsilon \sim n^{4/3}$	$a \sim t^{1/2}$	$1 < t < M^{-2}$
stage 3 $\rbrace$	$\varepsilon = Mn$	$\lbrace a \sim t^{2/3}M^{1/3}$	$M^{-2} < t < MC^{-3/2}$
stage 4	$p \ll \varepsilon$	$a = C^{1/2}t$	$MC^{-3/2} < t$

During the first stage the gravitational interaction of neighboring particles is ~ 1 and has an important influence on the equation of state. An investigation of this stage is essential for the determination of the initial inhomogeneities of quantum character. If $C = 0$ then the fourth stage does not set in.

Investigations carried out by E. M. Lifshits [2] and Ya. B. Zel'dovich (verbal communication) of the laws governing the buildup of small density perturbations can be summarized as follows. Neglecting the effects of the spatial pressure gradient, the perturbations build up during all the stages in inverse proportion to $\dot{a}^2$,

$$z_{\varkappa}(t) = \frac{z_{0\varkappa}}{\dot{a}^2} \qquad (4)$$

where $z_{0\varkappa}$ is the "initial" amplitude at $\dot{a} \sim 1$.

Ya. B. Zel'dovich called our attention to the importance of taking into account the pressure effects $p \sim \varepsilon$ during the second stage for short waves. According to Zel'dovich, the stability condition is $M^{-2} > t > \varkappa^{-2}$ (that is, in particular, $\varkappa > M$). If the stability condition is satisfied, oscillations of the particle density take place with conservation of the adiabatic invariant.

When $\varkappa \gtrsim M$, formula (4) for the third and fourth stages should be replaced by a formula that takes into account the effect of the pressure gradient in the second stage:

$$z_{\varkappa}(t) = \frac{z_{0\varkappa}B(\varkappa)}{\dot{a}^2} \qquad t > M^{-2} \qquad (5)$$

Calculation of the function $B(\varkappa)$ for $\varkappa \sim M$ calls for the knowledge of the equation of state when $n \sim M^3 = 10^{42}$ baryon/cm^3.

We accept in this paper Zel'dovich's hypothesis that the initial temperature of matter in the universe was zero, and concentrate our attention on quantum fluctuations of the density.

We note that during the course of the expansion of the universe the temperature of the matter first increases somewhat, owing to the incomplete adiabaticity of the transformation of the elementary particles in the second stage, after which it drops as a result of the adiabatic expansion of matter, reaching $\sim 10^{-7}$ erg or less at densities on the order of 10^{42} baryon/cm^3. Zel'dovich's estimates (verbal communication) show that the temperature effect is not significant in the formation of inhomogeneities.

II. FUNDAMENTAL EQUATIONS FOR SMALL DENSITY PERTURBATIONS

The equations derived in this section coincide in content with the equations of E. M. Lifshits [2], but are written in a different coordinate system, with a view of going over to quantum theory. The quantization of small oscillations is easiest to carry out if the equations of motion are written in Lagrangian form with

$$\mathscr{L}_\varkappa = \tfrac{1}{2} m(t) \dot{z}^2 - \tfrac{1}{2} r(t) z^2$$

We then introduce a wave function $\psi_\varkappa (z,t)$, satisfying the Schrödinger equation [$m(t)$ and $r(t)$ are known functions]

$$-\frac{1}{2m(t)} \frac{\partial^2 \Psi}{\partial r^2} + \frac{r(t) z^2}{2} \Psi = -i \frac{\partial \Psi}{\partial t} \qquad (6)$$

The "mass" m should correspond here to the value (7), obtained for short-wave oscillations (that is, for an unperturbed metric) as a result of a simple direct calculation of the change in the fraction of the Lagrange function

$$\mathscr{L}_m = \int \sqrt{-g}\; \varepsilon\, d\xi$$

connected with the matter. Expanding in powers of z^2 and $\dot{z}^2$, we get

$$m_\varkappa(t) = \frac{a^2}{\varkappa^2} \frac{d\varepsilon}{dn}$$

$$r_\varkappa(t) = n \frac{d^2 \varepsilon}{dn^2} = \frac{dp}{dn} \qquad (7)$$

To obtain r in the case of long-wave perturbations, it is best to start from the equations of motion, obtaining the latter by a method which is close to that of Ya. B. Zel'dovich (verbal communication).

Let us consider the expansion (1) of an arbitrary perturbation in orthogonal functions Φ, satisfying the equation

$$\Delta \Phi + \left(\frac{\varkappa}{a}\right)^2 \Phi = 0 \qquad (8)$$

The function Φ can be written in the form of an exponential or a sinusoid, or else in spherically symmetrical form

$$\Phi \sim \frac{\sin \varkappa \xi}{\varkappa \xi}$$

(ξ is the radius). The dependence of $z_\varkappa$ on t is determined in all cases only by the value of the parameter $\varkappa$ in Eq. (8). To find this dependence, it is sufficient to satisfy the equations of motion at the point $\xi = 0$ in the spherically symmetrical case. Then, by virtue of the isotropy of space, the equations of motion will be satisfied automatically in all the remaining points. The fundamental Eq. (2) is the equation of motion at the point $\xi = 0$, accurate to effects of the spatial pressure gradient.

For long-wave perturbations we can neglect the effects of the spatial pressure gradient. We shall denote by the index zero the unperturbed quantities and by the index 1 infinitesimally small quantities of first order:

$$a = a_0 + a_1$$
$$t = t_0 + t_1 \qquad (9)$$
$$\varepsilon = \varepsilon_0 + \frac{d\varepsilon}{da} a_1, \ldots$$

Substituting (9) in (2) and separating the terms of first order of smallness we obtain equations for the perturbation amplitude. Here we consider t as the local time [$dt = ds$ at the point $\xi = 0$; it is just this t which enters in (2)], and t_0 is regarded as the "world time" of the coordinate system.

Lifshits and Zel'dovich have assumed that $t_1 = 0$. Zel'dovich actually used variation of the integration constants in the analytic solution of (2), a simple matter, if one uses the special equation of state $\varepsilon \sim n^\gamma$; he considered $\gamma = 1$, $\gamma = 4/3$, and $\gamma = 2$. In the general case, the substitution (9) with the condition $t_1 = 0$ in (2) leads to the equation of motion

$$\frac{1}{a_0} \frac{d}{dt}\left(a_0^2 \frac{d}{dt}\left(\frac{a_1}{a_0}\right)\right) = 4\pi \left(n \frac{d\varepsilon}{dn} + 3n \frac{dp}{dn}\right) a_1 \qquad (10)$$

We note that inasmuch as one of the solutions of this equation should be $a_1 \backsim \dot{a}_0$ [a time shift of the unperturbed solution of (2)], Eq. (10) can be written in the form

$$\frac{\ddot{a}_1}{a_1} = \frac{\ddot{a}_0}{\dot{a}_0} \qquad (11)$$

Introducing the relative amplitude of the inhomogeneities

COSMOLOGY

$$z = \frac{n_1}{n_0} = -\frac{3a_1}{a_0} \tag{12}$$

we obtain Zel'dovich's result from (10) or (11) for the special equation of state $\varepsilon \sim n^\gamma$, namely, two independent solutions

$$z \sim t^\alpha$$

$$\alpha_1 = 2 - \frac{4}{3\gamma}$$

$$\alpha_2 = -1$$

We now forgo the condition $t_1 = 0$, imposing an additional condition which is useful to us, $m = (a/\varkappa)^2 \, d\varepsilon/dn$ [as in (7)]. The first term in an equation of the type (10) should be of the form ($\omega \equiv d\varepsilon/dn$)

$$\frac{1}{a_0} \frac{d}{dt_0}\left(\omega a_0^2 \frac{d}{dt_0}\left(\frac{a_1}{a_0}\right)\right)$$

Obviously, t_1 must be chosen so as to add to the equation a term

$$a_0 \frac{d\omega_0}{dt} \frac{d}{dt}\left(\frac{a_1}{a_0}\right)$$

We put

$$\dot{t}_1 = -\frac{1}{\omega}\frac{d\omega}{dt}\left(\frac{da_0}{dt}\right)^{-1}$$

$$a_1 = -z\frac{dp}{d\varepsilon} \tag{13}$$

Recognizing that, accurate to second-order terms,

$$\frac{d^2 a}{dt^2} = \ddot{a}_0 + \ddot{a}_1 - 2\dot{t}_1 \ddot{a}_0 - \ddot{t}_1 \dot{a}_0$$

we obtain under the gauge condition (13):

$$\frac{1}{a^2 \omega}\frac{d}{dt}\left(\omega a_0^2 \frac{dz}{dt}\right) = \left[4\pi n \left(\frac{d\varepsilon}{dn} + 3\frac{dp}{dn}\right) \right.$$

$$\left. - 8\pi \frac{dp}{d\varepsilon}(\varepsilon + 3p) + 3\frac{\dot{a}_0}{a_0}\frac{d}{dt}\left(\frac{dp}{d\varepsilon}\right) \right] z \tag{14}$$

[Starting with (14) we shall omit the subscript zero in differentiating with respect to the world time t_0.]

For the special equation of state $\varepsilon = n^\gamma$ we obtain again two linearly independent solutions:

$$z \sim t^\alpha$$

$$\alpha_1 = 2 - \frac{4}{3\gamma}$$

$$\alpha_2 = 1 - \frac{2}{\gamma}$$

The second solution vanishes (for long-wave perturbations) when the reference frame is transformed to $t_1 = \text{const}$. To the contrary, when $t_1 = -z\, dp/d\varepsilon$, the solution with $\alpha_2 = -1$ vanishes.

For the short-wave oscillations, Eq. (14) should be supplemented by a term which takes into account the spatial pressure gradient [according to formula (7)]. Taking (3) into account, the equation takes (for the particular case $C = 0$) the form

$$\frac{1}{a^2\omega} \frac{d}{dt}\left(a^2\omega \frac{dz}{dt}\right) = \left[4\pi n \left(\frac{d\varepsilon}{dn} + 3\frac{dp}{dn}\right) - 8\pi \frac{dp}{d\varepsilon}(\varepsilon + 3p) \right.$$

$$\left. - 24\pi\varepsilon n \frac{d}{dn}\left(\frac{dp}{d\varepsilon}\right) - \frac{\varkappa^2}{a^2} \frac{dp}{d\varepsilon} \right] z \quad (15)$$

When $\gamma = \text{const}$, the solution of this equation is expressed in terms of Bessel functions; for example, when $\gamma = 4/3$ we have increasing and decreasing solutions of the form ($\vartheta \sim t^{1/2}\varkappa$)

$$z \sim \begin{cases} \cos\vartheta - \vartheta^{-1}\sin\vartheta \\ \sin\vartheta + \vartheta^{-1}\cos\vartheta \end{cases}$$

III. THE FUNCTION $B(\varkappa)$ AND THE MASSES OF STARS OF PREGALACTIC ORIGIN

Yu. M. Shustov and V. A. Tarasov have at our request solved Eq. (15), with the aid of an electronic computer, for different values of $\varkappa$. The calculations were made for the simplest equation of state, satisfying $\varepsilon = nM$ with $n^{1/3} \ll M$ and $\varepsilon = An^{4/3}$ with $n^{1/3} \gg M$ (A is a constant ~ 1):

$$\varepsilon = n(M^2 + A^2 n^{2/3})^{1/2} \quad (16)$$

The asymptotic value of p as $n \to 0$ is less satisfactory in this equation;

for $n^{1/3} > M$ no account was taken of the transformations and interactions of the baryons.

The function $a(t)$ can be obtained in the case of Eq. (16) analytically (Shustov). Shustov and Tarasov find, by integrating (15), the limiting value as $t \to \infty$ of the auxiliary variable

$$\zeta = z\left(1 + \frac{a^2 M^2}{A^2}\right)^{-1/2}$$

putting $d\zeta/dt = dz/dt \sim z_0$ as $t \to 0$. It is obvious that $\zeta(\infty) \backsim z_0 B$.

In accordance with the results of the sections that follow, we put $z_0 \sim \varkappa$. Also, $\zeta(\infty)$ is a function of the parameter $A^{1/2}\varkappa$. This function is oscillating and sign alternating, but attenuates rapidly with increasing $\varkappa$.

The distribution of the stars over the masses depends in a complicated nonlinear manner on z, and can hardly be obtained without very cumbersome calculations. For a qualitative estimate let us determine the function $F(N) = \Delta N/N$, where N is the average number of particles in a certain volume $V = N/n$ and ΔN is the mean-square deviation:

$$(\Delta N)^2 = \overline{[N(V) - \bar{N}(V)]^2}$$

We obtain $F(N)$ from formulas which, according to N. A. Dmitriev (verbal communication from Zel'dovich), can be employed when $F(N)$ is of such form that it is essential to exclude the effect of the fluctuations on the boundary of the averaging region. In our notation

$$F(N) = \left(\frac{\int_0^\infty z^2(\varkappa)\varkappa^2 e^{-\varkappa^2/4\alpha^2}\, d\varkappa}{N\int_0^\infty \varkappa^2 e^{-\varkappa^2/4\alpha^2}\, d\varkappa}\right)^{1/2} \quad (17)$$

$$N = 4\pi \int_0^\infty e^{-\alpha^2 \xi^2} \xi^2\, d\xi \quad \text{i.e., } \alpha = \frac{\pi^{1/2}}{N^{1/3}}$$

The intuitive meaning of formulas (17) is that ΔN and N are calculated for a volume V with diffuse boundaries. For statistically independent particles we have $z = 1$ and $F(N) = N^{-1/2}$.

It is obvious that the formation of objects with N particles is possible only if $F(N) \sim 1$. It can be assumed that if there exist maxima of the plot of the distribution of the stars by mass, then they correspond at least approximately to the maxima of functions of the form $N^m F(N)$ ($M \sim 1$). Putting (arbitrarily) for estimating purposes $m = \frac{1}{2}$, we obtain, using the results of Shustov and Tarasov (assuming $A = 5.8$, corresponding to a mixture of 1 proton + 1 electron + 1 neutrino without account of the transformation of the baryons and of the interactions) a maximum of the function $N^{1/2}F$ at $N = 0.47 \times 10^{57}$, that is, at a mass $m = 0.4 M_\odot$. The assumption that the stars preferably produced are those with mass $\sim M_\odot$ is due to Zel'dovich, who started from an examination of the boundaries of the stability region.

One could have assumed that the stars which enter in globular clusters were produced at an earlier stage of the expansion of the universe, and their mass distribution should be described by the theory developed above. Sandage [3] investigated the distribution of the stars of the globular cluster M3 with respect to their absolute stellar magnitudes. Maxima were obtained near $0.8 M_\odot$ and at $(3-4) M_\odot$. Whereas the first maximum could be related somehow with the foregoing theory, after refining the equation of state and the averaging laws, the presence of the second mass maximum (which consists essentially of the variable stars of type RR-Lyrae), cannot be explained. It is obvious that stars brighter than the sun cannot be of the same age as the universe, and at least part of their mass is of secondary origin.

The function $F(N)$ for arbitrary $z(x)$ increases monotonically with decreasing N (remark by Shustov). It is possible that this means that most of the primary stars have mass smaller than $0.4 M_\odot$ (determined by phenomena which we did not take into account), and all the masses of Sandage's distribution are of secondary origin.

IV. THE HYPOTHESIS CONCERNING THE EQUATION OF STATE WITH $n \sim 1$

In the case of a degenerate Fermi gas of relativistic noninteracting particles, $\varepsilon = A n^{4/3}$ with $A \sim 1$. The electromagnetic interaction of the extremely relativistic fermions in all orders of the perturbation theory series is $\sim n^{4/3}$, that is, it only changes the value of the coefficient A. We assume that an account of the mutual transformation of fermions and of all possible interactions still leaves us with a formula of the type

$$\varepsilon = A(n) n^{4/3} \quad \text{where } A(n) \sim 1 \text{ for } 1 > n^{1/3} > M$$

Another assumption of the theory is that we can extrapolate the concepts of general relativity down to distance scales of the order of 1.61×10^{-33} cm. In all probability, at such scales our concepts of space should be reviewed, that is, we are on the border of applicability of the existing theory, a fact which does not exclude the possibility of obtaining qualitatively correct results in the sense of the "correspondence principle."

At densities of the order of unity (in gravitational units) the exchange and correlation gravitational interactions of fermions become comparable in order of magnitude with the Fermi energy. Therefore the gravitational interactions become decisive in the equation of state. Qualitatively the role of these effects is obvious from the expansion of ε in powers of the gravitational constant [more accurately, in powers of $(G\hbar/c^3)n^{2/3}$]:

$$\varepsilon = An^{4/3} + Bn^2 - Cn^{8/3} \tag{18}$$

In this expansion, which is valid when $n \ll 1$, the second term is the exchange gravitational interaction, and the third is the correlation interaction. A, B, and C are of the order of unity and are larger than zero. When $n \sim 1$ the kinetic and exchange energies are as before greater than zero, and the correlation energy is as before less than zero, but owing to the effect of the subtraction of quantitatively unknown expressions, any theoretical estimates of even the qualitative nature of the curve $\varepsilon(n)$ are hypothetical. We assume that a dependence analogous to (18) takes place also for $n \sim 1$, that is, for a certain value $n = n_\alpha$ the quantity ε reaches a maximum, and $\varepsilon = 0$ when $n = n_0$ (see Fig. 1, curve a).

An equation of state of this form makes it possible to formulate correctly the problem of calculating the initial perturbations (see below).

The qualitative character of the unperturbed solution is also more satisfactory. The dependence of the world radius a on the time is obtained from (2). For $t = 0$ we put $\dot{a} = 0$, i.e.,

$$\varepsilon = \frac{-3C}{8\pi a^2}$$

$$\ddot{a} \sim -(\varepsilon + 3p) \approx -3n\frac{d\varepsilon}{dn} > 0$$

For $T \ll 1$ we have $a - a_0 \sim t^2$. Also, $\varepsilon + 3p = 0$ at the point of inflection of the $a(t)$ curve, and beyond $a \sim t^{1/2}$.

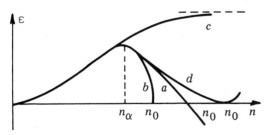

Figure 1

The solution for $a(t)$ can be symmetrically continued into the region $t < 0$:

$$a(t) = a(-t) \tag{19}$$

It is natural to assume that this symmetry extends to all physical properties, so that

$$\Psi(t) = \Psi^*(-t) \tag{20}$$

where Ψ is the state vector. The entropy of a certain part of the universe is $S(t) = S(-t)$, $S(0) = 0$. Such a "doubling" of physical reality is by virtue of its identical character not very meaningful, nor does it lead to any difficulties in principle. But on the other hand this eliminates the customarily raised question: "And what when $t < 0$?"

We shall also consider in what follows a variant of the equation of state (a plot of which is shown in Fig. 1, curve b), for which $d\varepsilon/dn \to \infty$ as $\varepsilon \to 0$. If we assume in this limiting case that $\varepsilon \sim (a - a_0)^{1/2}$, then we can readily obtain from (3) the dependence of $a - a_0$ on t (it is obviously logical to consider in this equation only the case $C = 0$):

$$a - a_0 \sim |t|^{4/3} \tag{21}$$

In this case, as in the case of curve a, it is logical to assume symmetry of the states under time reversal. An alternate solution of the problem of negative time is possible (for $C = 0$), if the equation of state has the form shown on curve c ($\varepsilon \to$ const as $n \to \infty$) or on curve d [$\varepsilon \to A(a - a_0)^2$ as $a \to a_0$]. In the case c, at the initial stage of the expansion of the universe, so long as $\varepsilon \approx$ const, $p = -\varepsilon$ and $\ddot{a}/a =$ const > 0, i.e.,

$$a = e^{\lambda t} \quad \text{as } t \to -\infty \tag{22}$$

Subsequently, when $\varepsilon \neq$ const ($\varepsilon \sim n^{4/3}$), the exponential growth is transformed into a growth like $a \sim t^{1/2}$. Analogously, for case d,

$$a - a_0 \sim e^{\lambda t} \quad \text{as } t \to -\infty$$

We can attempt to determine the true character of the equation of state at $n \sim 1$ by comparing the observational data with the deductions of the theory concerning the magnitude of the initial density inhomogeneities.

V. QUANTUM THEORY OF OCCURRENCE OF INITIAL PERTURBATIONS AT $n \sim 1$

Small density perturbations can be described with the aid of a wave function that depends on the dynamically independent generalized "normal" coordinates $z_\varkappa$. In the case of statistical independence this function is of the form

$$\Psi(t, z_1, z_2, \ldots) = \prod_\varkappa \Psi_\varkappa(z_\varkappa, t) \qquad (23)$$

Each of the functions $\Psi_\varkappa{}^\dagger$ is described by its own Schrödinger equation of the harmonic oscillator type (24), with the "mass" $m(t)$ and "elasticity" $r(t)$ as variables (it is not necessary here to have $m > 0$ and $r > 0$):

$$-\frac{1}{2m}\frac{\partial^2 \Psi}{\partial z^2} + \frac{rz^2}{2}\Psi = -i\frac{\partial \Psi}{\partial t} \qquad (24)$$

Here

$$m = \frac{a^2}{\varkappa^2}\frac{d\varepsilon}{dn} \qquad (25)$$

$$r = -\frac{a^2}{\varkappa^2}\frac{d\varepsilon}{dn}\left[4\pi n\left(\frac{d\varepsilon}{dn} + 3\frac{dp}{dn}\right) - 8\pi\frac{dp}{d\varepsilon}(\varepsilon + 3p)\right.$$

$$\left. + 3\frac{\dot a}{a}\frac{d}{dt}\left(\frac{dp}{d\varepsilon}\right)\right] + n\frac{d^2\varepsilon}{dn^2} \qquad (26)$$

The Schrödinger equation with a potential energy proportional to z^2 has two classes of self-similar solutions:

$$\Psi_+(z, t) = v_+(t)e^{-\mu(t)z^2} \qquad (27)$$

$$\Psi_-(z, t) = v_-(t)e^{-\mu(t)z^2}z \qquad (28)$$

These solutions satisfy (24) if the complex parameters μ, v_+, and v_- satisfy the following ordinary differential equations:

$$\frac{d\mu}{dt} = i\left(\frac{2\mu^2}{m} - \frac{r}{2}\right) \qquad (29)$$

$$\frac{dv_+}{dt} = \frac{i\mu v_+}{m} \qquad (30)$$

$$\frac{dv_-}{dt} = \frac{3i\mu v_-}{m} \qquad (31)$$

The initial values of v and μ can be arbitrary (with Re $\mu > 0$).

We note that the general solution of the Schrödinger equation can be represented in the form of a contour integral with respect to the parameter μ. For example,

---
$\dagger$We shall henceforth omit the index $\varkappa$ in most cases.

$$\Psi(z, t) = \frac{1}{2\pi}\int_{-\infty}^{+\infty} dy_0 \, e^{-\mu(t)z^2} [C_+(y_0)v_+(t) + zC_-(y_0)v_-(t)] \qquad (32)$$

At the initial instant of time we have here $\mu(0) = x_0 + iy_0, x_0 = \text{const} > 0$, and the time-independent functions C_+ and C_- are determined from the integrals

$$v_+(0)C_+(y_0) = \int_{-\infty}^{+\infty} dz \, |z| \, e^{-\mu(0)z^2}\Psi(z, 0) \qquad (33)$$

and analogously for C_-. A similar representation in the form of multiple integrals can be written for the elements of the Landau–Neumann density operator (the case of a "mixture").

We assume that each of the degrees of freedom of $z_\varkappa$ is described by a wave function of the type (27)

$$\Psi_+ = v_+(t)e^{-\mu(t)z^2}$$

We put below

$$\mu = x + iy = \varkappa^{-2}(X + iY) \qquad (34)$$

Obviously

$$\overline{z^2} = \frac{1}{4x} \qquad (35)$$

or (in order of magnitude)

$$z \sim \varkappa X^{-1/2} \qquad (36)$$

We omit in (29) the small term $n(d^2\varepsilon/dn^2) = dp/dn$ from formula (26) for r [we confine ourselves to the case of "long" waves; this is possible if the first term in $r \backsim (a/\varkappa)^2 (d\varepsilon/dn)$ is not anomalously small]. We find that the equation for $\varkappa^2\mu = X + iY$ no longer contains $\varkappa$. We arrive at the conclusion that if the solution of Eq. (29) exists at all, then the initial inhomogeneity $z_0(\varkappa)$ should be proportional to $\varkappa$ and should not contain other parameters which differ from unity in order of magnitude. Thus, for long-wave perturbations,

$$z \sim \frac{\varkappa}{\dot{a}^2} \qquad (37)$$

Using the law of averaging (17), we obtain

$$F(N) = \frac{\Delta N}{N} \sim \frac{N^{-5/6}}{\dot{a}^2} \qquad (38)$$

(with a proportionality factor ~ 1).

For $N \lesssim M^{-3}$ we need a correction for the pressure effect in the second stage:

$$F \sim \frac{N^{-5/6} B(N)}{\dot{a}^2} \tag{38a}$$

where

$$[B(N)]^2 = \frac{\int_0^\infty B^2(\varkappa) z_0^2 \varkappa^2 e^{-\varkappa^2/4\alpha^2} d\varkappa}{\int_0^\infty z_0^2 \varkappa^2 e^{-\varkappa^2/4\alpha^2} d\varkappa}$$

At the present time $\dot{a}^2 = 10^{-52} - 10^{-53}$. Therefore, the formation of clusters with $F(N) \sim \Delta N/N \sim 1$ could have occurred up to the present time for masses containing $N = 10^{(6/5)52}$ or $10^{(6/5)53}$ particles, that is, masses of the order of $m = 3 \times 10^5 - 3 \times 10^6 M_\odot$. In order of magnitude this is the mass of a globular cluster.

The detailed discussion of the physical processes occurring when $n \sim 1$, contained in this paragraph, is of necessity beyond the level of our present-day knowledge. But even if our notions change very radically, the correctness of formula (38) is not excluded. A solution of (29) over the entire time interval undoubtedly exists if stability (that is, $r/m > 0$) corresponds to the initial instant of time $t \to 0$ or $t \to -\infty$. This criterion is satisfied by the equation of state represented by curve a of the figure ($r < 0$, $m < 0$), and is not satisfied by the equations of state of curves c and d. In these two cases, by virtue of the condition $d\varepsilon/dn \to 0$ as $n \to n_0$ or ∞, the principal term in the "elasticity" r is the term of the spatial pressure gradient dp/dn (short-wave term), which has in both cases a sign opposite to that of $d\varepsilon/dn \backsim m$. Therefore $dp/d\varepsilon < 0$, which leads to a short-wave instability in the initial period of time as $t \to -\infty$. But even if an account of the nonlinear terms ensures that "initial" perturbations are finite (which is doubtful), there is no doubt that the amplitude of the initial perturbations obtained in such theories is too large (gigantic clusters of galaxies will be produced). For these reasons we turn to curves a and b.

For a unique determination of $\mu(0)$ in the case of curve a we must stipulate in addition

$$\dot{\mu}(0) = 0 \tag{39}$$

In this case

$$\mu(0) = \frac{1}{2}(rm)^{1/2}$$

$$z(0) = \frac{1}{2\sqrt{2}} (rm)^{1/4} \tag{40}$$

However, the requirement that the wave function (39) be stationary for the nonstationary state ($\ddot{a} > 0$) does not seem quite logical to us. It is therefore possibly of interest to consider the equation of state of curve b. In this case

$$a - a_0 \sim t^{4/3}$$
$$m \sim -t^{2/3}\varkappa^{-2} \tag{41}$$
$$r \sim t^{8/3}\varkappa^{-2}$$

[the sign of r is determined by summation of all three terms in the square brackets of (26)]. Among the solutions of (29) there is one distinct solution that admits of a power-law approximation as $t \to 0$ (self-similar instability):

$$\mu = C_0 t^{-5/3} \varkappa^{-2} \tag{42}$$

Here C_0 is a definite constant ~ 1, Re $C_0 > 0$, and Re $\mu \to \infty$ as $t \to 0$; that is, with such an equation of state the matter in the universe experiences in the initial state no density fluctuations.

We now consider the behavior of the solution of the system of equations for X and Y [equivalent to Eq. (29)] as $t \to \infty$ and as $t \to t_\alpha$ [t_α is the instant when $m = (a/\varkappa)^2(d\varepsilon/dn)$ vanishes[†]]. Assuming that $\varepsilon \to n^{4/3}$ as $t \to \infty$, we then have an asymptotic solution

$$Y = \frac{2}{3}\left(\frac{32\pi}{3}\right)^{1/4} t^{-1/2}$$
$$X = B_\infty t^{-2} \tag{43}$$

The constant B_∞ can be arbitrary if the inequality $X \ll Y$ is satisfied. This asymptotic solution describes an increase in z proportional to t, corresponding to the result of the classical (that is, not quantum) theory. If the constant B_∞ does not depend on $\varkappa$, it is obvious that we have the already mentioned result (38), $\Delta N/N \sim N^{-5/6} \dot{a}^{-2}$. Near the point $t = t_\alpha$, the integral curve passes through an antinode-type singular point:

$$Y = Y_\alpha + A_\alpha(t - t_\alpha)^2$$
$$X = B_\alpha(t - t_\alpha)^2 \tag{44}$$

When $t < t_\alpha$ it is obvious that A_α and $B_\alpha \sim 1$, since $X_0 \sim 1$ as $t \to 0$ in

[†] E. B. Gliner (in a paper now in press) calls states with $d\varepsilon/dn = 0$, in connection with the isotropic character of the four-tensor $T^i_k \sim \delta^i_k$, a μ vacuum.

the case of an equation represented by curve a, or $C_0 \sim 1$ in the case of an equation represented by curve b. If analytic continuation of the solution in the vicinity of the point t_α is possible, than A_α and B_α are continuous and $B_\infty \sim 1$. However, the possibility of analytic continuation gives rise to certain doubts, for when $t \to t_\alpha$ we have $z \sim 1/(t - t_\alpha)$, that is, $z \to \infty$ (according to the linear theory). Actually it is necessary to take account near the point t_α of nonlinear effects.

The singular character of the point t_α can be greatly reduced by going over to a reference frame with $t_1 = 0$. This causes us to assume that a change in the order of magnitude of z at the point t_α is unlikely [although it is quite possible that when $t > t_\alpha$ the statistical state of each elementary "oscillator" z_x is no longer described by the wave function (27), but by a density operator].

VI. COSMOLOGICAL HYPOTHESIS

Let us assume that formula (38a) is correct. In this case the instant of formation of clusters containing N particles with mass $m = NM$ is determined by the condition $F(N, t) = \Delta N/N \sim 1$, that is, (taking into account that $\dot{a} \sim t^{-1/3} M^{1/3}$)

$$t(N) \sim N^{5/4} M [B(N)]^{-3/2} \sim 10^2 \text{ years} \times \left(\frac{m}{M_\odot}\right)^{5/4} B^{-3/2} \qquad (45)$$

The first to be formed are obviously "primordial" stars with mass $m \leq 0.4 M_\odot$ (at $t \sim 10^2$ years). From similarity considerations we assume that the fraction of the primordial gas not captured by the stars decreases in accordance with a power law of the form $[t(N_\odot)/t]^\alpha$. Under likely values of the exponent α, no more than several percent gas remains at a time corresponding to 10^9 years.

How do we explain the formation of galaxies [for which, according to (45), $t(N) \sim 10^{16}$ years]? The answer presented here for discussion contains several hypothetical assumptions.

Clusters containing not more than a certain critical number of primordial stars will sooner or later experience the gravitational collapse of Tolman–Oppenheimer–Snyder–Volkoff. (The question of the evolution of stellar clusters, leading to collapse, has recently been discussed in connection with the problem of quasars; see, for example, Ref. 4. The duration of the evolution decreases if a gaseous or dustlike phase is present in the cluster.)

An estimate shows that even after $t \sim 10^6$ years it is possible for quasars with a mass of 500 $M_\odot$, for which the sum of the formation time given by (45) and the time of hydrogen burnup are close to minimum and of the order of 10^6 years, to collapse. This is followed by collapses of larger clusters of matter. According to our hypothesis, which is apparently confirmed by the data on quasar observation, several percent of the matter is ejected during the collapses, with jet velocities of the order of 10^3-10^4 km/sec. (The ejection of gas clouds from M-82 confirms the lower of these figures.) The collapse results in a "post-collapse" object (PC object), which has very small dimensions and is manifest principally through its gravitational field.

It is possible that the bulk of the matter in the universe is contained at present in PC objects; with this assumption, an estimate of the average density of matter, in which only galaxies are taken into account, turns out to yield much too low a value and the disparity with the condition $C = 0$ disappears (it is precisely in this context that several authors have discussed this hypothesis; see, for example, the article by Zel'dovich [1] with a reference to I. D. Novikov). Incidentally, the intergalactic primordial stars with low luminosity, and intergalactic gas and dust of low density, are also very difficult to observe, and we do not exclude the possibility that it is just these components that constitute a principal or appreciable fraction.

We propose that the process of formation of galaxies began at $t \sim 10^9$ years and still continues, and will continue to infinity, encompassing ever increasing masses; the nonuniformity of the gas distribution is first produced in this case by jets due to collapses of clusters of primordial stars, and during the later stages an ever increasing role will be assumed by collapses of cores of galaxies and clusters of PC objects. By the time $t \sim 10^9$ years is reached the proposed fraction of the gas component is approximately 10%; these are remnants of the primordial gas and of gas produced by small collapses. At $t \sim 10^9$ years there occur collapses of clusters containing $10^5 M_\odot$. The kinetic energy of the jets ejected in such collapses can reach, according to our hypotheses, $10^{53}-10^{55}$ erg; this energy goes over into the energy of the shock waves propagating through the gas. Streams of gas are produced, and in regions where several streams meet the density of the gas increases by a factor of several times. As a result of the gravitational contraction of these condensations, galaxies are subsequently produced (possibly in pairs or in groups), with masses amounting to approximately $\frac{1}{3}-\frac{1}{5}$ of the mass of the gas. If the time for the formation of the condensations is 10^9 years, then the formation of a cluster with mass 10^{44} g requires a velocity of 10^6 cm/sec, that is, an energy of 10^{56} erg. According to our assumption, this is the energy of $10-10^3$ collapses of masses equal to $10^5 M_\odot$, or $1-10^2$ collapses of masses of $10^6 M_\odot$.

The gas clouds which become condensed in collisions of jets should possess in most cases an angular momentum and should constitute the flat component of the galaxies. To complete the picture, we must assume that, after the condensed regions of gas have been formed, some small fraction ($\sim 1\%$) of the globular clusters of primordial stars "left intact" by the collapses, and a similar fraction of the PC objects, is captured by the galaxies, thus forming the spherical components of galaxies. In accordance with the observations, the spherical component of a galaxy practically does not participate in the galactic rotation of the gas or in the formation of stars of secondary origin from the gas.

The mass of single PC objects (that is, not included in the clusters and in the core) in our galaxy is apparently much smaller than the mass of the galaxy, otherwise their gravitational field would affect the speed of galactic rotation. (This does not exclude a considerable number of PC satellites of the galaxy with former semiaxes.)

Globular clusters not captured by galaxies have not been observed. This may be due to their lack of bright and variable stars, characteristic of galactic clusters, which apparently arise (or are captured) when galactic clusters pass through the core of the galaxy.

Of course, the picture described is very hypothetical and must be supplemented, refined, or rejected as a result of further considerations.

The author is grateful to Ya. B. Zel'dovich for numerous discussions that led to formulation of the problem as a whole and enriched the work with many ideas. The author is also grateful to I. E. Tamm and E. M. Lifshits for a discussion and valuable remarks.

REFERENCES

1. V. A. Fock, *UFN 80*:353 (1963); Ya. B. Zel'dovich, *UFN 80*:357 (1963); E. M. Lifshits and I. M. Khalatnikov, *UFN 80*:391 (1963). *Sov. Phys. Uspekhi* 6:473 (1964), 6:475 (1964), and 6:495 (1964), trans.
2. E. M. Lifshits, *JETP 16*:987 (1946).
3. B. Bok and P. Bok. *Mlechnyĭ put'*, Fizmatgiz, 1959, p. 117. (*The Milky Way*, Harvard University Press, Cambridge, Mass., 1957.)
4. Ya. B. Zel'dovich and I. D. Novikov, *UFN 84*:377 (1964) and 86:447 (1965). *Sov. Phys. Uspekhi* 7:763 (1965) and 8:522 (1966), trans.

Paper 7

Violation of *CP* Invariance, *C* Asymmetry, and Baryon Asymmetry of the Universe

The theory of the expanding universe, which presupposes a superdense initial state of matter, apparently excludes the possibility of macroscopic separation of matter from antimatter; it must therefore be assumed that there are no antimatter bodies in nature, i.e., the universe is asymmetrical with respect to the number of particles and antiparticles (*C* asymmetry). In particular, the absence of antibaryons and the proposed absence of baryonic neutrinos implies a nonzero baryon charge (baryonic asymmetry). We wish to point out a possible explanation of *C* asymmetry in the hot model of the expanding universe (see Ref. 1) by making use of effects of *CP* invariance violation (see Ref. 2). To explain baryon asymmetry, we propose in addition an approximate character for the baryon conservation law.

We assume that the baryon and muon conservation laws are not absolute and should be unified into a "combined" baryon-muon charge $n_c = 3n_B - n_\mu$. We put

$n_\mu = -1, n_K = +1$ for antimuons μ_+ and $\nu_\mu = \mu_0$
$n_\mu = +1, n_K = -1$ for muons μ_- and $\nu_\mu = \mu_0$
$n_B = +1, n_K = +3$ for baryons P and N
$n_B = -1, n_K = -3$ for antibaryons P and N

This form of notation is connected with the quark concept; we ascribe to the p, n, and λ quarks $n_c = +1$, and to antiquarks, $n_c = -1$. The theory proposes that under laboratory conditions processes involving violation of n_B and n_μ play a negligible role, but they were very important during the earlier stage of the expansion of the universe.

Source: Нарушение СР-инвариантности, С-асимметрия и барионная асимметрия Вселенной, Письма в ЖЭТФ 5:32-35 (1967). Reprinted from *JETP Lett.* 5: 24–27 (1967), with permission from American Institute of Physics.

We assume that the universe is neutral with respect to the conserved charges (lepton, electric, and combined), but C asymmetrical during the given instant of its development (the positive lepton charge is concentrated in the electrons and the negative lepton charge in the excess of antineutrinos over the neutrinos; the positive electric charge is concentrated in the protons and the negative in the electrons; the positive combined charge is concentrated in the baryons, and the negative in the excess of μ neutrinos over μ antineutrinos).

According to our hypothesis, the occurrence of C asymmetry is the consequence of violation of CP invariance in the nonstationary expansion of the hot universe during the superdense stage, as manifest in the difference between the partial probabilities of the charge-conjugate reactions. This effect has not yet been observed experimentally, but its existence is theoretically undisputed (the first concrete example, Σ_+ and Σ_- decay, was pointed out by S. Okubo as early as 1958) and should, in our opinion, have much cosmological significance.

We assume that the asymmetry has occurred in an earlier stage of the expansion, in which the particle, energy, and entropy densities, the Hubble constant, and the temperatures were of the order of unity in gravitational units (in conventional units the particle and energy densities were $n \sim 10^{98}$ cm^{-3} and $\varepsilon \sim 10^{114}$ erg/cm^3).

M. A. Markov (see Ref. 3) proposed that during the early stages there existed particles with maximum mass of the order of one gravitational unit ($M_0 = 2 \times 10^{-5}$ g in ordinary units), and called them maximons. The presence of such particles leads unavoidably to strong violation of thermodynamic equilibrium. We can visualize that neutral spinless maximons (or photons) are produced at $t < 0$ from contracting matter having an excess of antiquarks, that they pass "one through the other" at the instant $t = 0$ when the density is infinite, and decay with an excess of quarks when $t > 0$, realizing total CPT symmetry of the universe. All the phenomena at $t < 0$ are assumed in this hypothesis to be CPT reflections of the phenomena at $t > 0$. We note that in the cold model CPT reflection is impossible and only T and TP reflections are kinematically possible. TP reflection was considered by Milne, and T reflection by the author; according to modern notions, such a reflection is dynamically impossible because of violation of TP and T invariance.

We regard maximons as particles whose energy per particle ε/n depends implicitly on the average particle density n. If we assume that $\varepsilon/n \sim n^{-1/3}$, then ε/n is proportional to the interaction energy of two "neighboring" maximons $(\varepsilon/n)^2 n^{1/3}$ (cf. the arguments in Ref. 4). Then $\varepsilon \sim n^{2/3}$ and $R_0^0 \sim (\varepsilon + 3p) = 0$, i.e., the average distance between maximons is $n^{-1/3} \sim t$. Such dynamics are in good agreement with the concept of *CPT* reflection at the point $t = 0$.

We are unable at present to estimate theoretically the magnitude of the *C* asymmetry, which apparently (for the neutrino) amounts to about $(\bar{\nu} - \nu)/(\bar{\nu} + \nu) \sim 10^{-8}$–$10^{-10}$.

The strong violation of the baryon charge during the superdense state and the fact that the baryons are stable in practice do not contradict each other. Let us consider a concrete model. We introduce interactions of two types.

1. An interaction between the quark-muon transformation current and the vector boson field $a_{i\alpha}$, to which we ascribe a fractional electric charge $\alpha = \pm 1/3, \pm 2/3, \pm 4/3$, and a mass $m_a \sim (10-10^3)m_p$. This interaction produces reactions $q \to a + \bar{\mu}$, $q + \mu \to a$, etc. The interaction of the first type conserves the fractional part of the electric charge and therefore the actual number of quarks minus the number of antiquarks ($= 3n_B$) is conserved in processes that include the a boson only virtually.

We estimate the constant of this interaction at $g_a = 137^{-3/2}$, from the following considerations: The vector interaction of the a boson with the μ neutrino leads to the presence of a certain rest mass in the latter. The upper bound of the mass μ_0 is estimated in Ref. 5 on the basis of cosmological considerations. If we assume a flat cosmological model of the universe and assume that the greater part of its density $\rho \sim 1.2 \times 10^{-29}$ g/cm^3 should be ascribed to μ_0, then the rest mass of μ_0 is close to 30 eV. The given value of g_a follows then from the hypothetical formula

$$\frac{m_{\mu_0}}{m_e} = \frac{g_a^2}{e^2} \sim 137^{-2}$$

We note that the presence in the universe of a large number of μ_0 with finite rest mass should lead to a number of very important cosmological consequences.

2. The baryon charge is violated if the interaction described in item 1 is supplemented with a three-boson interaction leading to virtual processes of the type $a_{\alpha_1} + a_{\alpha_2} + a_{\alpha_3} \to 0$. According to B. L. Ioffe, I. Yu. Kobzarev, and L. B. Okun', the Lagrangian of this interaction is assumed to be dependent on the derivatives of the a field, for example,

$$\mathcal{L}_2 = g_2(\sum_\alpha f_k^i f_j^k f_i^j + \text{h.c.}) \qquad f_{ik} = R_0 \partial a_i$$

Inasmuch as $\mathcal{L}_2$ vanishes when two tensors coincide, in this concrete form of the theory we should assume the presence of several types of a fields. Assuming $g_2 = 1/M_0^2$ and $M_0 = 2 \times 10^{-5}$ g, we have strong

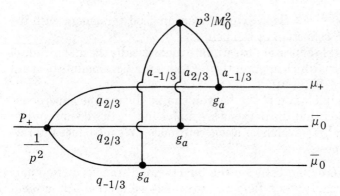

Figure 1

interaction at $n \sim 10^{98}$ cm^{-3} and very weak interaction under laboratory conditions. Figure 1 shows a proton-decay diagram including three vertices of the first type, one vertex of the second, and the vertex of proton decay into quarks, which we assume to contain the factor $1/p_q^2$ (due, for example, to the propagator of the "diquark" boson binding the quarks in the baryon). Cutting off the logarithmic divergence at $p_q = M_0$, we find the decay probability

$$\omega \sim \frac{m_p^5 \, g_a^6 \, [\ln(M_0/m_a)]^2}{M_0^4}$$

The lifetime of the proton turns out to be very large (more than 10^{50} years), albeit finite.

The author is grateful to Ya. B. Zel'dovich, B. Ya. Zel'dovich, B. L. Ioffe, I. Yu. Kobzarev, L. B. Okun', and I. E. Tamm for discussions and advice.

REFERENCES

1. Ya. B. Zel'dovich, *UFN* 89:647 (1966) (review); *Sov. Phys. Uspekhi* 9:602 (1967), trans.
2. L. B. Okun', *UFN* 89:603 (1966) (review); *Sov. Phys. Uspekhi* 9:574 (1967), trans.
3. M. A. Markov, *JETP* 51:878 (1966); *Sov. Phys. JETP* 24:504 (1967), trans.
4. A. D. Sakharov, *ZhETF Pis'ma.* 3:439 (1966); *JETP Lett.* 3:288 (1966), trans. [Paper 13 in this volume].
5. Ya. B. Zel'dovich and S. S. Gershtein, *ZhETF Pis'ma* 4:174 (1966); *JETP Lett.* 4:120 (1966), trans.

Paper 8

Quark-Muonic Currents and Violation of *CP* Invariance

In Ref. 1 we postulated, from cosmological considerations, the existence of quark-muonic currents whose interaction constant g_a with the fractional-charge vector field $a_{i\alpha}$ is of the order of $137^{-3/2}$. In this note we consider a hypothesis in which the violation of *CP* invariance in $K_{0,L}$ decay (see Ref. 2) is ascribed to the difference between the phase constants $g_a \exp(i\phi)$ for ordinary and strange quarks.

We assume interaction Lagrangians with maximum phase difference for ordinary and strange quarks[†] and with *P*-parity conservation:

$$\mathscr{L} = \sum_{\alpha,q,\mu} g_a[(\bar{\Psi}_{-q} a_{i\alpha} \gamma^i \Psi_\mu) + \text{h.c.}]$$

$$\mathscr{L} = \sum_{a,\mu} g_a \lambda[(\bar{\Psi}_{-\lambda} a_{i\alpha} \gamma^i \Psi_\mu) - \text{h.c.}] \qquad (1)$$

$$\alpha + q + \mu = \alpha + \lambda + \mu = 0$$

Here q, λ, μ, and α are the indices of the electric charge and take on the values $q = -\frac{1}{3}, +\frac{2}{3}$; $\lambda = -\frac{1}{3}$; $\mu = 0, 1, 0$; $\alpha = -\frac{2}{3}, +\frac{1}{3}, +\frac{4}{3}$.

Generally speaking, the constant g_a can depend on the index α, but we shall not consider these variants.

Figure 1 shows the main diagrams for the transformation of $K_0 = \lambda \bar{n}$ into $\bar{K}_0 = \lambda \bar{n}$. The matrix element of the transition, $V_{12} = (K_0|V|\bar{K}_0)$, is complex:

$$V_{12} \sim 2g_W^2 g_{W\lambda}^2 + i g_W g_{W\lambda} g_a g_{a\lambda}$$

Source: Кварк-мюонные токи и нарушение СР- инвариантности, Письма в ЖЭТФ 5:36-39 (1967). Reprinted from *JETP Lett.* 5:24−27 (1967), with permission from American Institute of Physics.

[†]The equivalent form of the theory—introduction of complex phases in the expression for the weak current $j_W = \bar{e}Ov + \bar{\mu}O\mu_0 e^{i\phi_1} + \bar{n}Ope^{i\phi_2} + (g_{W\lambda}/g_W)\bar{\lambda}Ope^{i\phi_3} + \cdots$, with $\phi_3 - \phi_2 = \pi/2$.

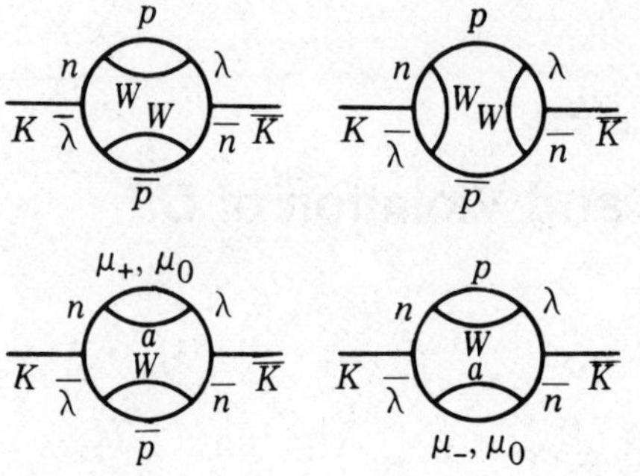

Figure 1

g_W is the constant for the interaction of the weak current with the W boson,

$$\frac{4\pi g_W^2}{m_W^2} = \frac{G}{\sqrt{2}} = \frac{10^{-5}}{\sqrt{2m_p^2}}$$

where m_W is the W-boson mass. In the expression for V_{12} we neglected the possible differences between the masses m_a and m_W, which enter in diverging expressions. The eigenfunctions of the mass operator are proportional to

$$K_O V_{12} \pm \bar{K}_O |V_{12}|$$

and, in particular,

$$K_L = \cos \nu \, K_2 + i \sin \nu \, K_1$$

where

$$\nu = \frac{1}{2} \frac{\operatorname{Im} V_{12}}{\operatorname{Re} V_{12}} = \frac{1}{4} \frac{g_a g_{a\lambda}}{g_W g_{W\lambda}}$$

The difference between K_L and K_2 determines completely the amplitude of the decay $K_L \to \pi^+ \pi^-$, since the "direct" decay $K_2 \to \pi^+ \pi^-$ is forbidden by P-parity conservation in the a interaction. Therefore

$$\nu = \left| \frac{A(K_L \to \pi^+ \pi^-)}{A(K_2 \to \pi^+ \pi^-)} \right| = 2 \times 10^{-3}$$

Using the value $g_a^2 = 137^{-3}$ from Ref. 1, we get $g_W^2 = 137^{-2}$ and

$$m_W \sim 10 m_P \sim 137 \frac{m_\pi}{2} = 137^2 m_e$$

m_a remains unknown.

The $q\mu$ currents, as well as the quark-electronic currents which are possible in principle, should change the ratio of the yields of the two channels of $\pi^{\pm}$-meson decay:

$$R = \left| \frac{A(\pi^+ \to e^+ \nu)}{A(\pi^+ \to \mu^+ \mu^0)} \right|^2$$

The experimental value is $R = (1.24 \pm 0.03) \times 10^{-4}$ (see Ref. 3) and agrees within the limits of measurement accuracy both with the theoretical value of R_W given by the V–A theory with electromagnetic corrections (see Ref. 4), and with the possible value of R_{W+a} measured as a result of the presence of $q\mu$ currents $(R_{W+a} - R_W)/R_W \sim \pm 0.01(m_W^2/m_a^2)$. We use the formula

$$R_W = \left(\frac{m_e}{m_\mu}\right)^2 \left(\frac{m_\pi^2 - m_e^2}{m_\pi^2 - m_\mu^2}\right)^2 \frac{f_e^2}{f_\mu^2} = 1.21 \times 10^{-4}$$

f takes into account the electromagnetic correction to the nonrelativistic approximation, as an effect that is due to the attraction of the charged particles in the neutral channel and depends on their relative velocity v:

$$f = \left|\frac{\Psi(0)}{\Psi(\infty)}\right| = \left\{\frac{2\pi e^2}{v[1 - \exp(-2\pi e^2/v)]}\right\}^{1/2}$$

$$\frac{f_e^2}{f_\mu^2} = 0.945$$

The change in the decay amplitude due to the $q\mu$ currents is

$$A_a = -A_w \frac{g_a^2}{2g_W^2} \frac{m_W^2}{m_a^2} \frac{m_\pi}{m_\mu} e^{i\phi}$$

(see Fig. 2). The uncertainty in the phase is here a reflection of the uncertainty in the relative phases in the expression (2) for the weak current. Putting $\phi_1 = \phi_2$ we get

$$R_{W+a} = R_W \left(1 - \frac{2m_\pi}{m_\mu} \frac{m_W^2}{m_a^2} v\right)^{-2}$$

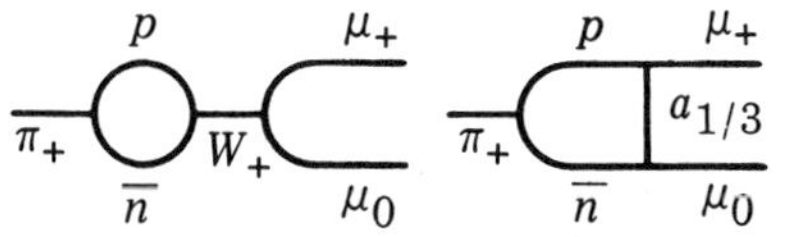

Figure 2

We note that the quark-electronic currents of comparable magnitude (for $m_a \sim m_W$ and $\phi_2 \neq \pi/2$) would change R by several times 10%, which is completely excluded by the experiment.

In the case of $K^\pm$ decay, a similar effect of the change in the yield ratio of the two channels is very small ($\sim v^2$) when the phases ϕ coincide, owing to the fact that the a amplitude is imaginary. An effect of the order of $v(m_W^2/m_a^2)$ (i.e., $\sim 1-0.1\%$) must be expected in the expression for the probability of K capture of μ^- in hydrogen and He^3, and in effects of transverse polarization in three-particle decay of K_L.

Particular interest attaches to effects of violation of C symmetry of the partial probabilities in decays with P-parity conservation and change of strangeness, for example, the deviation of the ratio of the partial probabilities

$$\frac{K^+ \to \pi^+ + \pi^+ + \pi^-}{K^- \to \pi^- + \pi^- + \pi^+}$$

$$\frac{\Sigma^+ \to N + \mu^+}{\Sigma^- \to N + \pi^-}$$

from unity (the Okubo effect). These effects are also of the order of $v(m_W^2/m_a^2)$, but since they depend on the phase differences $\phi_I \sim A_a/A_W$ for different values of the isospin I and on the phases of the strong interaction, their numerical value should be much lower than 0.1%.

As indicated by L. B. Okun' in the course of a discussion, processes that can be used to test the theory are those in which pairs of charged mesons are produced, for example, $K^+ \to \pi^+ + \mu^+ + \mu^-$ (relative yield $\sim v^2$). The processes $K_L \to \mu^+ + \mu^-$ and $K_L \to \mu^+ + \mu^- + \pi^0$ are strongly forbidden, but the processes $K_L \to \mu^+ + \mu^- + \gamma$ and $\Sigma^+ + P^+ + \mu^+ + \mu^-$ are possible (all are $\sim v^2$).

The author takes this opportunity to thank L. B. Okun' for discussion and advice.

REFERENCES

1. A. D. Sakharov, *ZhEFT Pis'ma* 5:32 (1967); *JETP Lett.* 5:24 (1967), trans. [Paper 7 in this volume].
2. L. B. Okun', *UFN* 89:603 (1966); *Sov. Phys. Uspekhi* 9:574 (1967), trans.
3. A. H. Rosenfeld et al., *Revs. Modern Phys.* 37:633 (1965) and UCRL-8030, part I, August 1965.

Paper 9

Antiquarks in the Universe

The purpose of this paper is to call attention to the possibility, on the assumption of a baryon-neutral and charge-asymmetric universe, that there exist in nature (cosmic and terrestrial) hypothetical particles with the following properties.

1. The particle has baryon charge $n_b = -\frac{1}{3}$; that is, it is an antiquark. The particle is stable.
2. The average cosmic concentration of antiquarks in the universe is three times the cosmic density of nucleons (the universe is baryon neutral and charge asymmetric in the absence of quarks and antinucleons).
3. The mass of the antiquark is 10 to 15 times that of the proton.
4. Stable antiquarks are assumed electrically neutral. Thus we are adopting a quark model with integral charge quarks. We also assume that the antiquark does not form bound states (diquark) with nucleons, and therefore is not captured by nuclei. These two assumptions remove manifest contradictions with experiment which would arise otherwise.
5. The antiquark is a hadron. Its cross section for scattering by nuclei is of the order of 10^{-24}–10^{-26} cm^2.

The assumption that $m = 10$–$15 m_P$ for the mass of the antiquark is based on the following premises. The mean density of matter in the universe is assumed to be "critical," i.e., related to the Hubble constant H by the equation

Source: Антикварки во Вселенной, в сборнике «Проблемы теоретической физики» посвященном 60-летию Н. Н. Боголюбова, «Наука», Москва, 35-44, 1969. (*Problems in Theoretical Physics,* Festschrift for the 60th birthday of N. N. Bogolyubov, Nauka, Moscow, pp. 35–44, translated by W. H. Furry.)

$$\rho_{\text{crit}} = \frac{3H^2}{8\pi G} = 2 \times 10^{-29} \text{ g/cm}^3$$

where G is the gravitational constant. As is well known (cf., e.g., Ref. 1) it follows from this relation that the spatial curvature of the universe is zero and that the asymptotic behavior of the expansion (a scale) is critical:

$$\dot{a} \to 0 \quad a \to \infty \quad \text{as } t \to \infty$$

We ascribe the greater part of the density of matter to the antiquarks distributed in intergalactic space, and the greater part of the density of nucleons to the mean cosmic density of the matter concentrated in galaxies. According to Oort's estimate, the latter quantity is $\rho_{\text{gal}} = 5 \times 10^{-31}$ g/cm^3. Using our assumed ratio 3:1 of the numbers of antiquarks to that of nucleons, we have for the ratio of the antiquark and proton masses

$$\frac{m}{m_P} = \frac{\rho_{\text{crit}}}{3\rho_{\text{gal}}} \approx 13$$

In this paper the existence of particles with the indicated properties is connected with the hypothesis of cosmological CPT symmetry of the universe [2] and with the three-triplet quark model[†] [3–11]. Experimental searches for such particles are justified, since their existence does not contradict known experimental evidence.

I. THE BARYON ASYMMETRY OF THE UNIVERSE

Quantum field theory assumes the CPT invariance of the equations. Along with this, the world around us does not contain any macroscopic numbers of antinucleons and positrons, so that we face an obvious puzzle. At the present time there are various (at least three) points of view with regard to this problem.

1. There is no baryon asymmetry on the cosmic scale; 50% of the galaxies consist of antimatter with negative baryon charge (for an exposition of this point of view see Ref. 12).

2. The baryon charge is a rigorously conserved quantity with a positive definite average value in the universe. In particular this view is accepted by the authors of Ref. 1. According to this concept, at the initial instant of the expansion of the hot universe, at temperatures much

---

[†] One of the originators of the three-triplet quark model, who has played an important part in advancing these hypotheses is Nikolaĭ Nikolaevich Bogolyubov.

larger (in units $\hbar = c = k = 1$) than the mass of the proton, there occurred a fractional "initial" excess of the number of baryons over the number of antibaryons

$$\frac{N_0 - \bar{N}_0}{N_0 + \bar{N}_0} \sim 10^{-9} - 10^{-8}$$

which was conserved after the annihilation, so that then

$$N_1 = N_0 - \bar{N}_0$$
$$\bar{N}_1 = 0$$

The violation of CP invariance in the decay of the K_{02} meson, discovered in 1964, is the basis of the third hypothesis, which was put forward in Ref. 2.

Already in 1958, S. Okubo showed that the violation of CP invariance in principle leads to a difference between the probabilities of charge-conjugate reactions (he gave the example of the ratios of the two decay channels for Σ_+ and the corresponding antiparticle; at the same time he pointed out that because of CPT invariance there is exact charge symmetry of the masses and the total decay probabilities). There are now specific experimental examples of charge asymmetries of partial probabilities. Owing to this "Okubo effect," under the nonstationary conditions of the expanding universe a charge asymmetry could arise with charge symmetric initial conditions. A logical completion and at the same time a justification of the idea of the charge symmetry of the initial state of the universe is the hypothesis, advanced in Ref. 2, that there is an identical CPT inversion of all natural phenomena relative to the singular hypersurface of zero extent in the Friedmann model of the universe.[†] This hypothesis answers in the most natural way the question of what there was before the instant of maximum density. The hypothesis is compatible with the CPT invariance of the equations of quantum field theory. It implies that the entropy increases with increase not of t, but of $|t|$, which may arouse doubts.

We note that in statistical mechanics the increase of entropy is a

---

[†]The cosmological CPT inversion is defined for the metric $ds^2 = dt^2 - a^2(dx^2 + dy^2 + dz^2)$ as a reflection in the hypersurface of zero extent $t = 0$, $a(0) = 0$, with the replacements $t \to -t$, $a \to -a$, particle $\to$ antiparticle, with x, y, z unchanged. In the hypothesis of Ref. 2 this is an identical transformation. According to arguments given in Ref. 2, in the neighborhood of the singularity $t = 0$ we have $a \sim t$ and $\varepsilon \sim S^{2/3}$ (ε is the energy density and S the entropy density). With such an equation of state the temperature $T = d\varepsilon/dS$ goes to zero as $S \to \infty$, and is of the same order as the interaction energy and quantum fluctuation per single particle. Therefore it is not excluded that at the initial instant $t = 0$ the matter was in a superfluid state, which would affect the spectrum of initial nonuniformities of the cosmological matter distribution.

consequence of the statistical boundary conditions. It is usually assumed that these conditions are imposed at $t \to -\infty$. In a model with a singular point it is most natural to impose boundary conditions in the neighborhood of this point as $t \to 0$, which leads to the indicated law of entropy increase.

To deal with the question of the baryon asymmetry it is necessary to make a specific assumption about the conservation of baryon charge. In Ref. 2 the writer proposed that the conservation of baryon charge was approximate in nature.[†] In the present paper we go back to the concept of exact conservation of the baryon charge.

3. We assume that the mean cosmic density of the baryon charge is always equal to zero. At the present stage of the existence of the universe the positive baryon charge is concentrated in nucleons, and the negative, in neutral stable antiquarks. This charge asymmetry arose at early stages of the expansion of the hot universe owing to the charge asymmetry of the partial probabilities in the nonstationary processes of expansion and cooling of the charge-symmetric plasma, with a gigantic entropy density and a temperature of the order of 10^{19} eV (as estimated below). Neutral (and other integral-charge) quarks were postulated independently by a number of writers in the three-triplet quark model proposed in 1964–1965 (see papers by Nambu [7], Han [8], Bogolyubov, Struminskiĭ, and Tavkhelidze [6], Morpurgo [9], and others). This model is essential for our argument.

II. THE THREE-TRIPLET MODEL

There are two difficulties with a one-triplet quark model: (a) The quarks with fractional charges proposed in it are not observed experimentally; (b) owing to the Fermi statistics the three quarks making up a baryon cannot have a symmetric coordinate wave function, which is the most natural dynamically. Both of these difficulties are removed in the three-triplet model [3–11]. We confine ourselves to the version of the model shown in Table 1. The nine different quarks will be characterized by a "column index" i taking the values $i = n, p, \lambda$, and a "row index" j taking the values $j = a, b, c$.

The frame in the lower right corner contains the values of the electric charge; the strangeness S and a new quantum number $\varkappa$ (the

---

[†] The assumption made in Ref. 2 about the nature of a violation of baryon conservation is not the only possible one. For example, by ascribing baryon charge 2 to the muon and the muon neutrino we could describe the appearance of the baryon asymmetry and at the same time maintain the stability of the proton and the practical stability of complex nuclei (only reactions involving four baryons would be possible).

COSMOLOGY

TABLE 1

j \ i		l_p	l_n	λ
	S \ κ	0	0	−1
α		1	0	−1	−1
a — b		0	1	0	0
a — c		0	1	0	0

"charm" charge) which we introduce are shown in borders of the frame. As in the one-triplet scheme, the λ quark in each row is assumed to differ strongly from the two "ordinary" quarks n,p in the same row. We use for the "ordinary" quarks the collective designation l quarks, $l = 1,2$.

Analogously, we assume that the α quark in each column has a mass very different from those of the quarks in the other two rows of the same column. We shall later give arguments in favor of choosing the first row. Of course, we also do not exclude differences, on a smaller scale, of the mass and other properties of the quarks with the indices b and c; in any case they are regarded as different in the sense of the Pauli principle. For the quarks b and c we use the collective designation a quarks. The important difference of α from a justifies the introduction of the quantum number κ, in analogy with the strangeness S of the λ quark, which differs from the l quarks.

The baryon charge is $n_b = +1/3$ for all nine quarks. For the nine antiquarks all the charges (n_b, electric charge, κ, and S) have the opposite sign.

The wave function of a baryon is assumed to be antisymmetric in the row index, which is equivalent to assuming a strong attraction between quarks with different indices j:

$$\psi = \frac{1}{\sqrt{6}} \begin{vmatrix} \alpha(1) & b(1) & c(1) \\ \alpha(2) & b(2) & c(2) \\ \alpha(3) & b(3) & c(3) \end{vmatrix}$$

With this assumption a symmetric coordinate S function and a symmetric spin–unitary-spin function are consistent with Fermi statistics. The "average" electric charge for quasiparticles p, n, λ is the same as in the one-triplet model; the consequences of that model are preserved for magnetic moments and other linear quantities.

The wave function for mesons is taken in the form

$$\psi = A\alpha\bar{\alpha} + Bb\bar{b} + Cc\bar{c}$$

(A, B, C are mixing parameters). This sort of structure must exist when there is a strong attraction between quarks and antiquarks of the same row j.

In Refs. 3 to 11 it is postulated that for pseudoscalar mesons $B = C = 0$ and $A = 1$ and for vector mesons $A = B = C = 1/3$. We shall start from a dynamical model and assume that the mass of α quarks is much less than that of a quarks. The absence of appreciable vector mixing of pairs (as evidenced in the ρ-ω degeneracy) means that the wave function of vector mesons must be of the form $\psi = \alpha\bar{\alpha}$ (in the lowest nonet).

For pseudoscalar mesons there is considerable mixing of strangeness, which leads, as is well known, to the mass difference of π_0 and η_0. We shall assume, however, that the mass difference between a quarks and α quarks is much larger than that between λ and l quarks; in this case the formula $\psi = \alpha\bar{\alpha}$ can also hold approximately for pseudoscalar mesons. In such a dynamical model one does not have the difficulty with meson states that are degenerate with respect to charm, which is inherent in some formulations of the three-triplet scheme. The masses of states orthogonal to the observed mesons form, according to these assumptions, a high-lying doublet $\psi \approx b\bar{b} \pm c\bar{c}$. Along with this, such a three-triplet scheme with the same composition for all mesons preserves a number of consequences of the one-triplet scheme, for example, the result that the magnetic matrix element of the decay $\omega_0 \to \pi_0 + \gamma$ and the magnetic moment of the proton are equal (for a survey of the one-triplet model see Ref. 13). By taking for the mesons the function $\psi = \alpha\bar{\alpha}$, we uniquely determine the choice of the charm row. As is well known, the mass of the charged K meson is smaller than that of the neutral K meson, from which we can conclude that the mass of the $p\alpha$ quark is smaller than that of the $n\alpha$ quark. On the other hand, we assume that the residual charmed antiquarks remaining in the cosmos are neutral and stable, i.e., have less mass than all other quarks. It follows from this that the first row with the charges $(0, -1, -1)$ is the charm row, and that the residual antiquarks are of the $p\alpha$ type.

III. THE KINETICS OF THE ORIGIN OF THE CHARGE ASYMMETRY

A measure of the charge asymmetry that the theory has to explain is the ratio of the mean cosmic density of baryons $n_B = 3 \times 10^{-7}$ cm^{-3} to the mean number of residual photons $n_\gamma = 5 \times 10^2$ cm^{-3} (cf. Ref. 1). This ratio, $\xi = 10^{-9}$ in order of magnitude, is the ratio of the final number of baryons to the initial entropy of the universe,

$$\xi = \frac{\lim_{t\to\infty} a^3 n_B}{\lim_{t\to 0} a^3 S} \sim 10^{-9}$$

where a is the scale of the metric and S is the entropy density. We shall hereafter use for estimates the gravitational system of units,

$$G = \hbar = c = k = 1$$

With the relativistic equation of state the expansion of the universe leads to the following formulas for the scale a, the temperature T, the entropy density S, and the density n of all relativistic particles for which $T \gg m$ (t is the age of the universe):

$$a \sim t^{1/2}$$

$$T \sim \frac{1}{a} \sim t^{-1/2}$$

$$n \sim S \sim T^3 \sim a^{-3} \sim t^{-3/2}$$

For nonrelativistic particles with $T \lesssim m$, $t \gtrsim 1/m^2$, the equilibrium density is much smaller than S; in gravitational units the mass of the proton is $m_p = 10^{-19}$ and the mass of the quark is $m = 10^{-18}$, so that the "critical instant" for quarks is $t_q = 10^{36}$ (10^{-7} sec in ordinary units). Our estimate for the reaction time of strongly interacting particles is

$$\tau \sim \frac{1}{n\sigma} \sim \frac{1}{T} \sim t^{1/2}$$

Using $H \approx 1/t$, we have for the characteristic "parameter of nonstationarity" for quarks, $\eta \approx H\tau \approx \tau/t \sim t^{-1/2} = 10^{-18}$. The fact that this parameter is small compared with ξ shows that the origin of the charge asymmetry must be referred to a time much earlier than $t_q = 10^{-7}$ sec, probably to a time $t_\kappa \sim 10^{18} = 10^{-25}$ sec or even earlier, when $\eta \gtrsim 10^{-9} = \xi$. We assume that owing to nonequilibrium processes at times $t \lesssim t_\kappa$ an asymmetry with respect to charm appears, and that during the further expansion this asymmetry is "frozen in"; accordingly, the interaction violating κ invariance does not have to be very strong. Actually it must be weaker than the Fermi interaction.

Let us consider the mixture of quarks and antiquarks formed up to the time $t = t_q$, with nonvanishing charm charge $\kappa \neq 0$, $n_B = 0$. Let N_α, $N_{\bar\alpha}$, N_a, $N_{\bar a}$ be the numbers of quarks and antiquarks of various types:

$$\begin{aligned} N_\alpha &= N_0 - \Delta & N_a &= N_1 + \Delta \\ N_{\bar\alpha} &= N_0 + \Delta & N_{\bar a} &= N_1 - \Delta \end{aligned} \Bigg\} \Delta \text{ mixture}$$

We have $\kappa = -2\Delta$, $n_B = 0$.

Let us consider reactions conserving κ. When the temperature falls below $T = m$ the quarks annihilate and the mixture goes almost entirely over into a low-temperature stable state, with total number N_B of baryons B and $N_{\bar{\alpha}} = 3\Delta$ (with the same values of κ and n_b). This state is stable against charm-conserving reactions; the reaction $\bar{\alpha} + \bar{\alpha} \to B + a$ is forbidden by charm conservation, and the reaction $\bar{\alpha} + B$ is forbidden by our hypothesis. The limiting state is not fully reached, since the density of matter and the number of collisions falls during the expansion; the remaining number of "relic" quarks $a = b$ and c is determined by the reaction

$$b + c \to B + \bar{\alpha}$$

The relative number of residual a quarks in ratio to photons is estimated as $\eta = \tau/t = 10^{-18}$ (i.e., it amounts to a fraction 10^{-9} of a number of baryons).

The remaining numbers of α and $\bar{a}$ quarks are determined by reactions with the $\bar{\alpha}$ quarks, which are relatively numerous, and so the numbers of residual α and $\bar{a}$ quarks are exponentially small. These arguments are analogous to the estimates in Ref. 14, which dealt, in particular, with the binary interaction of two quarks.[†]

We ascribe the origin of the κ asymmetry to nonequilibrium CP-noninvariant processes at temperatures $T > 1/t_\kappa^{1/2} = 10^{-9}$ eV. There have been no studies of elementary processes at such energies. For illustrative purposes we confine ourselves to a consideration of a "model" which perhaps reflects some qualitative features of reality. We consider six states with the quantum numbers (S is strangeness) shown in Table 2 (see Fig. 1). All these states are neutral in the electric and baryon charges and can be converted into each other. Of the possible transitions we consider only those shown in the diagram. States with these quan-

TABLE 2

	$\varkappa$	S
A	+1	0
B	+1	+1
C	+1	−1
$\bar{A}$	−1	0
$\bar{B}$	−1	−1
$\bar{C}$	−1	+1

[†]The writer is grateful to Ya. B. Zel'dovich for a fruitful discussion of these questions.

COSMOLOGY

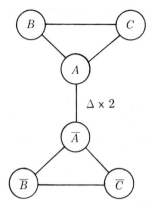

Figure 1

tum numbers can be realized with two quarks; for example, the state B has the same quantum numbers as the pair of quarks αl and $\alpha \lambda$. It is understood that we are dealing with conditions in a dense plasma at $T \sim 10^{19}$.

The occupation numbers of these states satisfy kinetic equations of the type

$$\frac{dA}{dt} = -A(\omega_{A\bar{A}} + \omega_{AB} + \omega_{AC}) + \bar{A}\omega_{AA} + B\omega_{BA} + C\omega_{CA}$$

(we consider the usual form for $t > t_0 > 0$), and so on, six equations in all. The transition probabilities ω_{ik} are subject to relations which follow from CPT invariance, but are otherwise arbitrary. We assume that there is no T invariance, i.e., $\omega_{ij} \neq \omega_{ji}$ (except that $\omega_{A\bar{A}} = \omega_{\bar{A}A}$), and that $\omega_{AB} \neq \omega_{AC}$. In this general case all the probabilities in the "triangles," which are subject to eight independent conditions, can be expressed in terms of four parameters; $\omega_{[AB]}$, $\omega_{[BC]}$, $\omega_{[CA]}$, and δ, with

$$\omega_{AB} = \omega_{\bar{B}\bar{A}} = \omega_{[AB]} + \delta$$

$$\omega_{BA} = \omega_{\bar{A}\bar{B}} = \omega_{[AB]} - \delta$$

$$\omega_{BC} = \omega_{\bar{C}\bar{B}} = \omega_{[BC]} + \delta$$

and so on.

The conditions

$$\omega_{ik} = \omega_{\bar{k}\bar{i}}$$

$$\sum_k \omega_{ik} = \sum_k \omega_{\bar{i}\bar{k}}$$

are satisfied. Owing to these relations the kinetic equations have the trivial stationary equilibrium solution

$$A = \bar{A} = B = \bar{B} = C = \bar{C} = \text{const}$$

It is interesting, however, that nonequilibrium (nonstationary) solutions of this system of equations cannot be charge symmetric, even if the initial conditions were symmetric. At a time $t = t_0$ let us set $A = \bar{A} = n_1$, $B = \bar{B} = C = \bar{C} = n_2 \neq n_1$ (charge symmetry $\kappa = 0$). For this time t_0 we have

$$\frac{d^2}{dt^2}(a - \bar{A}) \neq 0$$

$$\frac{d^3\kappa}{dt^3} \neq 0$$

which illustrates the appearance of κ asymmetry for $t > t_0$. The arguments of this section are largely of an illustrative nature, based on a particular model. Their purpose is to show that there are no obvious contradictions between the proposed hypothesis and experimental knowledge.

IV. ANTIQUARKS IN NATURE AND THE POSSIBILITY OF OBSERVING THEM[†]

We suggest that all astronomical objects have originated from an approximately homogeneous medium of composition $B + 3\bar{\alpha}$ as results of a gravitational instability mechanism. Furthermore, we distinguish between objects of the first generation, which arise from a stationary medium with its particles $\bar{\alpha}$ (these objects are of the composition $B + 3\bar{\alpha}$) at a low temperature, and second-generation objects, which arise from a magnetically turbulent medium with large (non-Maxwellian) velocities of the particles $\bar{\alpha}$. These large velocities can be caused by statistical acceleration in variable gravitational fields. The composition of the second-generation objects, which we suppose include all or most galaxies, is almost purely of baryons ($\bar{\alpha}/B \ll 1$), since the particles $\bar{\alpha}$ do not take part in the magnetohydrodynamic and turbulent processes of transfer of angular momentum, which are important at this stage of the cosmological process, and since fast particles are not captured by a contracting object if their speed v satisfies the inequality $v^2/2 > |d\phi/dt|(L/v)$ (ϕ is the

---

[†]The writer is grateful to F. L. Shapiro, who originated the calculations and considerations presented in this section.

potential at the boundary of the object and L is its size). This is of course a crude preliminary picture and given here only as informal discussion.

As a preliminary basis for calculating estimates, we assume that the average density of $\bar{a}$ antiquarks is the same in our galaxy as its average in the universe, i.e., $n = 10^{-6}$ cm^{-3}, and their average velocity is of the same order as the velocity from galactic rotation, $v = 2 \times 10^7$ cm/sec. Of two possibilities for capture of antiquarks by the earth (during contraction of the dust cloud and during the subsequent time $T = S \times 10^9$ years) the second is most important. We estimate the total number of captured antiquarks as $2nvT\pi R^2 = 6 \times 10^{36}$. The captured antiquarks are in local temperature equilibrium with the terrestrial matter, and are not scattered in cosmic space, but distributed radially according to the formula of thermodiffusional equilibrium,

$$n_1(\overline{\sigma v})_{T_1} = n_0(\overline{\sigma v})_{T_0} \exp\left(-\frac{m}{k}\int_0^R dr \frac{g}{T}\right)$$

where $(\sigma v)_T$ is the average of the product of cross section and velocity at temperature T.

For our estimates we assume that the matter density varies linearly from $\rho_1 = 2.3$ at the surface to $\rho_0 = 15$ at the center, so that the difference of the gravitational potentials is

$$\phi(R) - \phi(0) = (\rho_0 + \rho_1)\frac{Rg}{\rho_0 + 3\rho_1} = 0.79Rg$$

We assume a constant temperature $T_0 = 3000$ K over almost the entire radius, falling to $T_1 = 300$ K in a thin layer of the earth's crust. We first set $m = 10m_p$ and find $n_0 = 10^{12}$ cm^{-3}, $n_1 = 3 \times 10^4$ cm^{-3}. When we set $m = 15m_p$, n_1 decreases sharply to 3 cm^{-3}. Evidently in planning a search for $\bar{a}$ antiquarks on the earth's surface one can assume initially that n_1 is between 1 cm^{-3} and 10^5 cm^{-3}. The density of a quarks at the center of the earth must be 10^9 times smaller, about 10^3 cm^{-3}, if at this density the burn-out owing to the reaction $b + c \to B + \bar{a}$ is still unimportant.

As Shapiro has remarked, on the basis of these estimates it seems feasible to perform experiments to observe antiquarks under laboratory conditions, in particular with scattering of ions by antiquarks. The most promising such experiment is probably the use of scattering of ions with energy ~1 MeV and mass $A \sim 10$ on the antiquarks in a vacuum chamber, with the recoil antiquarks registered outside the chamber (suggestion of Ya. B. Zel'dovich). Other possibilities are detection of recoil quarks inside the vacuum chamber, and a macroscopic experiment on the production of resonant torsional oscillations of a heavy rotor owing to transfer of angular momentum by antiquarks scattered from a nearby rotating mass.

The writer expresses his gratitude for discussions of theoretical and experimental questions to S. M. Vernov, N. M. Frank, L. B. Okun', V. B. Braginskiĭ, and all the members of the seminar of the Neutron Physics Laboratory of the Joint Center for Nuclear Research.

REFERENCES

1. Ya. B. Zel'dovich and I. D. Novikov, *Relyativistskaya astrofizika*, Nauka, Moscow (1967); *Relativistic Astrophysics*, Vol. I: *Stars and Relativity*, Vol. 2: *The Universe and Relativity*, University of Chicago Press, Chicago (1971), trans.
2. A. D. Sakharov, *ZhETF Pis'ma 5*:32 (1967); *JETP Lett. 5*:24 (1967), trans. [Paper 7 in this volume].
3. Z. Maki, *Prog. Theor. Phys. 31*:331 (1964).
4. Y. Hara, *Phys. Rev. 134*:B701 (1964).
5. H. Bacry, J. Nuyts, and L. Van Hove, *Phys. Lett. 9*:279 (1964).
6. N. N. Bogolyubov, B. V. Struminskiĭ, and A. N. Tavkhelidze, *Preprint*, Joint Inst. Nucl. Res. D-1968, 1965.
7. Y. Nambu, *Coral Gables Conference* (1965).
8. M. Y. Han and Y. Nambu, *Phys. Rev. 139*:B1006 (1965).
9. G. Morpurgo, *Physics 2*:95 (1965).
10. N. Cabbibo, L. Maiani, and G. Preparata, *Phys. Lett. 25B*:132 (1967).
11. J. Otokozawa and H. Suura, *Phys. Rev. Lett. 21*:1295 (1968).
12. H. Alfvén and O. Klein, *Arkiv for Fys. 23*:187 (1962). H. Alfvén, *Astrophys. J. 142*, No. 1 (1965).
13. E. M. Levin and L. L. Frankfurt, *Uspekhi Fiz. Nauk 94*:243 (1968); *Sov. Phys. Uspekhi 11*:106 (1968), trans.
14. Ya. B. Zel'dovich, L. B. Okun', and S. B. Pikel'ner, *Uspekhi Fiz. Nauk 87*:113 (1965); *Sov. Phys. Uspekhi 8*:702 (1966), trans.

Paper 10

A Multisheet Cosmological Model[†]

A multisheet cosmological model is considered. In it, there is an infinite sequence of four-dimensional continua with alternating general cosmological contraction and expansion; these are called sheets. The sheets are combined into a common time-ordered structure suggested by Novikov in which the metrics of local collapses of one sheet are continued to the region of anticollapses of the following sheet. The flat average spatial metric assumed for each sheet ensures the completeness and uniqueness of the "joining together" of the sheets. Six variants of the model are discussed.

A multisheet structure of the universe is proposed. It is based on ideas of Novikov [1,2] on the possibility of joining together by means of gravitational collapse two four-dimensional spaces, one of which is in the "absolute future" of the other. Novikov considered, in particular, the gravitational collapse of electrically charged dust; in this special case, there exists an explicit exact solution without true singularity (apart from the point at the center of the sphere) and the entire kinematics of the transition of the worldlines of the dust particles to the second space can be readily traced (see Fig. 1 and the Appendix). In the Appendix, the dynamics and kinematics of asymmetric collapse-anticollapse of matter with pressure are also considered. Here too there is a joining together of two spaces. We shall say that such spaces are joined together by collapse *sheets* and considered an infinite sequence of sheets joined pairwise, thus forming the general structure of the universe.

As a second basic assumption, the multisheet model of the universe

Source: Многолистная модель Вселенной, Препринт Института прикладной математики АН СССР, 1970. (Preprint, Institute of Applied Mathematics, Moscow, 1970.) Translated by J. B. Barbour.
[†]Dedicated to the memory of my wife, Claudia Alekseevna Vikhireva.

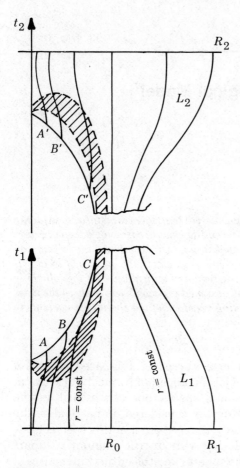

Figure 1 The joining together of sheets along the line ABC.

uses the limiting case of the Friedmann cosmological model corresponding to the euclidean spatial interval for the metric averaged over sufficiently large regions of space,

$$ds^2 = dt^2 - [a(t)]^2(dx^2 + dy^2 + dz^2) \tag{1}$$

where $a(t)$ is the scale factor. Its law of variation is given by the well-known formula [2]

$$\frac{da}{dt} = \pm \left(\frac{8\pi}{3} G\varepsilon a^2\right)^{1/2} \tag{2}$$

in which ε is the mean energy density, including gravitational energy. For dust, $\varepsilon \sim 1/a^3$, and

COSMOLOGY

$$a \sim |t|^{2/3} \tag{3}$$

A feature of the metric (1) is the unbounded growth in the limit as $a \to \infty$ of arbitrarily small initial perturbations of the density by an arbitrarily large factor (in contrast to the hyperbolic metric, for which perturbations grow only by a finite amount). Linear perturbation theory leads [3] to a change of $\Delta\varepsilon/\varepsilon$ in accordance with the law

$$\frac{\Delta\varepsilon}{\varepsilon}(x,y,z,t) = \frac{t_1}{t}\phi_1(x,y,z,t_1) + \left(\frac{t}{t_1}\right)^{2/3}\phi_2(x,y,z,t_1) \tag{4}$$

(we have restricted ourselves to the case of "dust").[†] The functions ϕ_1 and ϕ_2 are determined by the state at $t = t_1$ and, in the general case, are nonvanishing. The second term describes the unbounded growth of perturbations during the expansion $a \sim t^{2/3} \to \infty$ (and the first term describes the behavior during contraction). When the perturbations reach the value $\Delta\varepsilon/\varepsilon \sim 1$, collapse occurs in a finite proper time (in the absence of pressure). If we consider sufficiently large regions of space, for example, clusters of galaxies, we can certainly ignore pressure (which, of course, does not preclude gravitational collapse of much smaller objects as well). The collapse gives rise to "black holes," which again undergo collapse in a second stage, and so on ad infinitum. Writing

$$\varepsilon = \varepsilon_0 + \varepsilon_1 + \varepsilon_2 + \varepsilon_3 + \cdots$$

(ε_0 is the energy not involved in the collapse, ε_1 is the energy of the black holes of the first stage of the collapse, etc.), we have $\varepsilon_0/\varepsilon \to 0$, $\langle_1/\langle + 0$, etc. If we average our sufficiently large distances $L \gg L_1$, the metric (1) is always valid, but $L_1 \to \infty$ as $t \to \infty$. In the limit as $t \to \infty$, any two objects, however far apart they may be, will ultimately become part of one black hole. This picture is a consequence of the metric (1), and it uniquely determines the metric and other processes of the next sheet joined to the one under consideration. Suppose that in sheet $\mathscr{L}_i$ there is expansion, accompanied by collapses. Consider sheet $\mathscr{L}_{i+1}$, in which there are accordingly anticollapses. It is obvious that the time scale for $\mathscr{L}_{i+1}$ begins with $t = -\infty$, i.e., this sheet contracts. It appears reasonable (although I know of no completely rigorous proof) that the limiting metric (1) and (2) on the sheet $\mathscr{L}_i$ uniquely determines the limiting metric on the sheet $\mathscr{L}_{i+1}$ with reversed sign in Eq. (2). This assertion also applies to the replacement of contraction by expansion. This second case of joining together of sheets (joining together of spaces which are

---

[†]If $p \neq 0$, then (4) is replaced by $\Delta\varepsilon/\varepsilon = (b_1/a)\phi_1 + (b_2/a)\phi_2$ (for long-wavelength perturbations), where $b_1 = \dot{a}$ and b_2 are two linearly independent solutions of the second-order equation $\ddot{b}/b = \ddot{a}/a$ [a is a given function of t determined by (2)].

dense on the average) has been frequently considered. We mention here the papers of Novikov and Ne'eman [4], who (in connection with quasars) conjectured the existence in a contracting universe ("sheet" in our terminology) of "premature" collapses, which correspond to anticollapses of delayed cores in an expanding world. Obviously, this is the same process as we are proposing for joining spaces that on the average have a low density.

We now have everything we need to formulate the hypothesis. We assume that in both directions the universe is an infinite sequence of sheets joined together in pairs with alternating general cosmological contraction and expansion (alternatively, the sequence may be semifinite).

Particularizing the model of the universe, let us consider the possibility of a change in the direction of the arrow of time. The arrow of time (the direction of increasing entropy and irreversible processses) is determined by the initial statistical conditions of evolution of the universe. Logically, two variants are possible: the initial conditions are "specified" at $t = -\infty$ or the initial conditions are "specified" on a singular hypersurface. We assume that in the neighborhood of this hypersurface space is homogeneous and isotropic, i.e., this is a Friedmann-type singularity (F singularity). It is obvious that there can be only one F singularity on which the arrow of time changes its direction. the transition from sheet to sheet in all the remaining cases preserves the direction of the arrow of time in the anisotropic collapses-anticollapses, the dynamics of these collapses being entirely determined by the inhomogeneous dynamics of the previous sheets.

From this discussion, it follows that there are different possible variants of the multisheet cosmological model.

1. A neutral quasi-periodic model. In this model, we do not assume the existence of an F singularity, and the statistical conditions are imposed on sheet number $i \to -\infty$. In this model, the universe is assumed to be neutral with respect to the baryon charge, the two lepton charges, and the electric charge. It is obvious under such an assumption that nonequilibrium processes do not lead to an increase in the mean entropy per baryons or any other charge S, since the quotient S/Q is ∞ on all sheets, and the model can indeed be assumed to be quasi-periodic. The separation of particles and antiparticles in this model can in principle occur both through macroscopic mechanisms of the type proposed by Alfvén [5] as well as through the violation of CP invariance in nonequilibrium processes.

2. I have suggested earlier [6] that there exists and F singularity

with respect to which there is complete (cosmological) CPT symmetry. In this hypothesis, the occurrence of baryon asymmetry was attributed either to the nonconservation of baryon charge or the existence of hitherto unobserved carriers of baryon charge, for example, neutral antiquarks (with, of course, violation of CP). In Appendix II, the equation of state in the neighborhood of the point $t \to 0$ in an isotropic boson gas is discussed.

In combination with the multisheet hypothesis, we obtain the "CPT-symmetric multisheet model." In such a model, the sheets $\mathscr{L}_i$ and $\mathscr{L}_{-i}$ are CPT reflections of each other. In sheets $i = \pm1, \pm3, \pm5, \ldots$, there is general cosmological expansion, and in sheets $i = \pm2, \pm4, \ldots$, contraction. The sheets with i of small modulus carry a "memory" of the F singularity, and differ objectively in their astrophysics from the sheets with $i \to \pm\infty$, which go over asymptotically into the quasi-periodic regime of variant 1 (the variants 3, 4, 5, and 6 have the same asymptotic behavior).

3. We can retain the F singularity and simultaneously assume that the invariant baryon and lepton charges are nonzero. In this case, the F singularity does not correspond to any symmetry because of the violation of CP invariance. In particular, we can assume that in the neighborhood of the F singularity the entropy per baryon tends to zero (cold model).

4. Following Gliner and Al'tshuler [7], we can assume that at maximally high densities we have the equation of state (energy density) $\varepsilon \to$ const, (pressure) $p = -\varepsilon$. For this equation of state, $a \to 0$ as $T \to -\infty$, and the sequence of sheets is semi-infinite, beginning with a sheet having an F singularity. The model may be hot, neutral, or

5. Cold, like 3.

6. Zel'dovich has suggested that the present stage of general cosmological expansion of hot matter was preceded by contraction of cold homogeneous matter [2]. In the language of the multisheet model, we assume a semi-infinite sequence of sheets without F singularity, the "initial" sheet $\mathscr{L}_0$ being assumed to be a contraction sheet ($a \to \infty$ as $t \to -\infty$). The "initial" state at $t_0 = -\infty$ is characterized by an infinitely rarefied and infinitely cold gas of protons, electrons, and (perhaps) neutrinos distributed randomly without correlations in a flat x, y, z space.

In Fig. 2, this variant is shown schematically as an example, Lagrangian coordinates being plotted along the horizontal (the lines $x, y, z =$ const are tangent to a principal direction of the energy tensor).

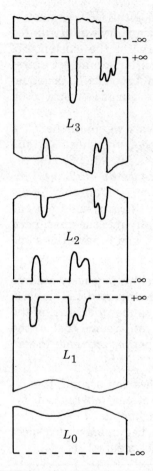

Figure 2 Variant 6. $L_0, L_2, L_4, \ldots$: contraction; $L_1, L_3, L_5, \ldots$: expansion.

The even and odd sheets correspond to cosmological contraction and expansion, respectively.

The symmetrically situated points in collapse on sheet i and in anticollapse on sheet $i + 1$ are identified.

CONCLUSIONS

A multisheet model of the universe and some variants of it have been proposed and in part justified [using Novikov's idea and the metric (1)]. The choice between these variants and the determination of the number of "our" sheet require more detailed comparison between theory and observations.

The hypothesis of the multisheet universe was strongly influenced by the work and ideas of I. D. Novikov, to whom I should like to express my deepest gratitude. It was a great help to be acquainted with the studies in Ref. 8 through the seminar of E. M. Lifshits and the preprint kindly supplied by him. A stimulus in the writing of this paper was a discussion with M. E.Gertsenshtein, who adopts an opposite point of view. I am grateful to Ya. B. Zel'dovich and the participants of the seminar at the Institute of Applied Mathematics for discussing this work.

APPENDIX I. COLLAPSE WITHOUT SINGULARITY

We explain Fig. 1, which shows the spherically symmetric collapse of charge dust on sheet 1 and anticollapse on sheet 2 (following Novikov [1]). In the figure, R is the Lagrangian radius, i.e., R = const for a dust particle, t is the proper time, and r is the Eulerian radius: $r = (S/4\pi)^{1/2}$, where S is the area of the sphere considered.

In the hatched regions, the lines r = const are spacelike; these are the so-called T regions [2]. For $R > R_0$, the dust particles are "strongly" charged and move apart; from $R < R_0$, they are "weakly" charged and undergo collapse; $R = R_0$ is the limiting line. The presence of the T regions means that the spaces 1 and 2 for $R > R_0$ are two different spaces but they are "joined together" along $ABC \leftrightarrow A'B'C'$ for $R < R_0$.

We now consider the collapse of a not completely symmetric mass of gas (which is a more general case then completely symmetric collapse); we assume that as $\varepsilon \to \infty$ the gas pressure does not vanish and satisfies the inequality $0 < \alpha_1 \leq p/\varepsilon \leq \alpha_2 < 1$.[†]

According to Zel'dovich [2], the case $p = \varepsilon$ is a limiting case, since it corresponds to a velocity of sound $(dp/d\varepsilon)^{1/2}$ equal to the velocity of light; we exclude this case.

Even small initial deviations from spherical symmetry grow during the collapse, and spherically or cylindrically symmetric collapse is an exceptional case (a rigorous mathematical analysis of this instability taking into account the effects of pressure has not yet been made).

Collapse of a "pancake" is possible (a two-dimensional case that is the only general case in accordance with our assumption). The gravitational field on the surface of a pancake of thickness Δ is of order of magnitude $\varepsilon\Delta$. The difference of the gravitational potentials between an interior and exterior point of the pancake is $\phi(\Delta/2) - \phi(0) \sim \varepsilon\Delta^2$ and tends to zero for the adopted equation of state as $\Delta \to 0$, i.e., locally, the

---

[†]The averaged gravitational field characteristics of a packet of gravitational waves (a geon in Wheeler's terminology) also satisfy these inequalities.

problem becomes Newtonian (or, more precisely, linear). The acceleration due to the gravitational force increases as $\omega_1 \to \varepsilon \Delta$, while the acceleration due to the pressure gradient increases as $\omega_2 = p/(\varepsilon + p)\Delta$, i.e., for the adopted equation of state the repulsion increases more rapidly than the attraction, and $\int(\omega_2 - \omega_1)\,d\Delta \to \infty$.

As a result, the contraction in the direction perpendicular to the plane of the pancake is replaced by expansion. In the other two directions, the contraction continues. The general kinematic picture of the contraction of a triaxial ellipsoid must consist of three successive "reflections" in three perpendicular planes (in the space of the lengths of the ellipsoid axes) with subsequent "transition" to sheet 2 if a T region exists and to sheet 1 otherwise. The part played by the T region has been discussed sufficiently in previous investigations, and we shall not dwell on it here. These considerations are an argument for the absence of a singularity of the matter density in asymmetric collapse. Of course, they need to be made more precise. Another case, which in a certain sense is intermediate between the two we have considered, is the collapse of a mass of gas prossessing pressure and a "frozen" dipole-type magnetic field. In this case, the contraction along two axes may be halted by the magnetic field, and the contraction along the third axis by the pressure gradient, as described above. Note that for asymmetric collapse we cannot speak of T reflection even when dissipative processes and shock waves are ignored.

We note finally that asymmetric anticollapse is accompanied by the emission of gravitational waves, whose fraction in the total energy balance can approach 100% for the limiting case of geons (Weber's experiments).

APPENDIX II. EQUATION OF STATE OF A BOSON GAS WITH ALLOWANCE FOR GRAVITATIONAL INTERACTION AND ITS BEARING ON THE HORIZON PROBLEM

In the CPT-symmetric model, as $t \to 0$ the quasiparticles of matter are truly neutral bosons with spin 0, which follows from CPT symmetry. We give arguments [6] that the limiting form of the connection between the entropy density and the energy density has the form $\varepsilon \sim S^{2/3}$.

We can take, for example, as our point of departure an ideal relativistic gas, for which $\varepsilon \sim S^{4/3}$. For a boson gas, S is not only the entropy density but also the particle density. As the particle density increases, the energy of each of the particles increases as $m = S^{1/3}$, and the energy of the gravitational interaction of two neighboring particles (i.e., at a distance $a \sim S^{-1/3}$) increases as $m^2/a \sim S$ (and it is negative, so that the particles attract each other). As a result, the equation of state with allowance for the gravitational correction has the form ($\alpha, \beta \sim 1$)

$$\varepsilon = \alpha S^{4/3} - \beta S^2$$

Here and below, we use gravitational units: $c = \hbar = G = 1$.

For $S = 1$ (in the same units, i.e., for $S = 10^{98}$ cm^3 in ordinary units) the two terms are comparable in magnitude. Assuming that $\varepsilon = mS$ in the limit $S \to \infty$ must have the same order as the energy $m^2 S^{4/3}$ of the exchange interaction, we find that in this limiting case $m \sim S^{-1/3}$ and finally $\varepsilon \sim S^{2/3}$. For such an equation of state $p = -\varepsilon/3$, and the dynamics of the expansion of the universe has the form $a \sim t$ [which follows for example, from (2) with $S \sim a^{-3}$]. Misner [8] has drawn attention to the horizon problem in the ordinary Friedmann model with equation of state $p > 0$, namely, the finite radius of the light cone when its apex (or base) tends to the point $t = 0$. A finite horizon cannot be readily reconciled with the observed isotropy of the microwave background. Misner regards this horizon problem as an argument against the isotropic cosmological model.

If the equation of state is $\varepsilon \sim S^{2/3}$, then $a \sim t$ and the Lagrangian radius of the horizon is

$$\int_{t_0}^{t_1} \frac{dt}{a} \to \infty \qquad \text{as } t_0 \to 0$$

i.e., the horizon problem is resolved without recourse to anisotropic models.

As we have indicated earlier, the resulting equation of state corresponds to a temperature $T = d\varepsilon/ds \to 0$ as $t \to 0$.

This equation does not preclude superfluidity of the initial state of the boson gas.

REFERENCES

1. I. D. Novikov, *ZhETF Pis'ma* 3:223 (1966); *JETP Lett.* 3:142 (1966), trans. *Ast. Zh.* 43:911 (1966); *Sov. Ast.* 10:731 (1967), trans.
2. Ya. B. Zel'dovich and I. D. Novikov, *Relyativistskaya astrofizika*, Nauka, Moscow (1967); *Relativistic Astrophysics*, Vol. I: *Stars and Relativity*, Vol. 2: *The Universe and Relativity*, University of Chicago Press, Chicago (1971), trans.
3. E. M. Lifshits, *ZhETF* 16:587 (1946). W. B. Bonnor, *Mon. Not.* 117:104 (1957). A. G. Doroshkevich and Ya. B. Zel'dovich, *Ast. Zh.* 40:807 (1963); *Sov. Ast.* 7:615 (1964), trans. A. D. Sakharov, *ZhETF* 49:345 (1965); *Sov. Phys. JETP* 22:241 (1966), trans. [Paper 6 in this volume].

4. I. D. Novikov, *Ast. Zh.* *41*:1075 (1964); *Sov. Ast.* *8*:857 (1965), trans. Y. Ne'eman, *Astrophys. J.* *141*:1303 (1965).
5. H. Alfvén, *Astrophys. J.* *142*:No. 1 (1965). H. Alfvén and O. Klein, *Ark. Fys.* *23*:187 (1962).
6. A. D. Sakharov, *ZhETF Pis'ma* *5*:32 (1967); *JETP Lett.* *5*:24 (1967), trans. *Preprint* R2-4267, JINR, Dubna [Paper 7 in this volume].
7. É. B. Gliner, *ZhETF* *49*:542 (1965); *Sov. Phys. JETP* *22*:378 (1966), trans. L. B. Al'tshuler, Oral communication, 1969.
8. E. M. Lifshits, Seminar at the P. N. Lebedev Physics Institute. V. A. Belinskii and I. M. Khalatnikov, *Preprints*, L. D. Landau Institute of Theoretical Physics. C. W. Misner, *Phys. Rev. Lett.* *22*:1071 (1969).

Paper 11

The Baryonic Asymmetry of the Universe

We discuss the possible appearance of an excess of baryons and antileptons during the early stage of expansion of a charge-symmetric hot universe in the framework of a unified gauge theory of strong, weak, and electromagnetic interactions. According to the estimates of the present paper, the baryon asymmetry $A = N_B/N_\gamma$ (the ratio of the mean baryon density to the density of quanta of the background radiation, which, up to a numerical factor, equals the ratio of the number of baryons to the initial entropy of the hot universe in the same comoving volume element) has the order of magnitude $A \sim \alpha^3 \vartheta^3 \delta_\alpha$ ($\alpha = g^2$ is the coupling constant of the gauge field, ϑ is a quantity of the order of the Cabibbo angle, δ_α is the phase of the complex quark mixing). The numerical coefficient in this formula may contain an additional small parameter. The paper presents some arguments relative to the "multifoliated" (many-sheeted) model of the universe previously proposed by the author.

I. INTRODUCTION. ESTIMATE OF THE EFFECT

In 1966 the author proposed the hypothesis of the appearance of an observable baryon asymmetry of the universe (and of a conjectured lepton asymmetry) during an early stage of the cosmological expansion from a charge-symmetric initial state. Such a process is possible owing to effects of *CP* violation under nonstationary expansion conditions, if one assumes nonconservation of the baryonic and leptonic charges [1].

In 1978 an analogous idea was formulated in a paper by Yoshimura [2]. Yoshimura indicates that in unified gauge theories of strong, weak, and electromagnetic interactions (cf. Ref. 3 and subsequent papers quoted in Ref. 2) baryon number is not conserved, owing to interactions

Source: Барионная асимметрия Вселенной, ЖЭТФ 76:1172-1181 (1979). Reprinted from *Sov. Phys. JETP* 49:594–599 (1979), translated by Meinhard E. Mayer, with permission from American Institute of Physics.

in which the "leptoquark" intermediate boson participates, and this together with the violation of CP invariance leads unavoidably to an excess of baryonic charge (baryon number) during the early stages of the expansion of the hot universe. Yoshimura indicates the possibility of a quantitative calculation of this effect by means of perturbation theory methods. While he was working on the present paper, the author also learned about the paper by Dimopoulos and Susskind [4] devoted to the same problem.

Below we obtain for the baryon asymmetry an estimate which is close to the one given by Dimopoulos and Susskind [4], but was obtained from a more detailed consideration of the kinetics of mutual transformation of particles and does not make use of the assumption that the mass of the leptoquark boson has the order of magnitude of the Planck mass $M_0 = 10^{19}$ GeV. Section V contains some considerations related to the multifoliated (many-sheeted) model of the universe proposed early by the author.

The other sections contain the reasoning behind the estimate of the baryon and lepton asymmetry. Here we summarize briefly the main points of this reasoning.

Deviations from particle-antiparticle symmetry manifest themselves only on account of the nonstationarity caused by the expansion of the universe. We denote the densities of different particle species by n_i. The equilibrium values of these densities will be denoted by n_i^0 and the deviations from the equilibrium situation will be characterized by the ratios

$$\frac{n_i'}{n_i^0} \qquad n_i' = n_i - n_i^0$$

The order of magnitude of the ratio n'/n^0 is $H\tau$, where τ is the characteristic reaction time of mutual transformations of particles, and H is the "Hubble parameter" characterizing the dynamics of the expansion of the universe. H is the logarithmic derivative of the "scale" a, the linear size of an arbitrary "comoving" volume element:

$$H = \frac{1}{a}\frac{da}{dt} = \left(\frac{8\pi}{3}G\rho\right)^{1/2} \tag{1}$$

ρ is the energy density and G is the gravitational constant. Here and in the sequel we have set $c = \hbar = k = 1$.

The quantity H is of the order of $1/t$, where t is the "age of the universe." Thus

$$\frac{n'}{n^0} \sim \frac{\tau}{t} \tag{2}$$

We assume that the most important processes of particle transformation are binary reactions with cross sections which tend to a constant in the high energy limit. With this assumption we have during the early stages of expansion of the universe

$$\tau \sim \frac{1}{n^0} \sim a^3$$

Since $a \sim t^{1/2}$ the relative deviations from the equilibrium value n'/n^0 are small and tend to zero as $t \to 0$.

For the appearance of an asymmetry it is essential that there exist a period of expansion of the universe when the temperature is of the order of the mass M_c of the leptoquark vector boson W_c which plays (cf. Sec. II and Refs. 2 and 3) a decisive role in the violation of baryon and lepton number conservation:

$$T_c \sim M_c$$

To this temperature corresponds a characteristic particle density $n_c^0 \sim M_c^3$, a characteristic energy density $\rho \sim M_c^4$, and according to Eq. (1), a characteristic age of the universe $t_c \sim H_c^{-1} \sim G^{-1/2} M_c^{-2}$, as well as a characteristic duration of the "decisive phase" (the length of the time interval which is most important for the processes we are interested in)

$$\Delta t_c \sim t_c$$

The violation of CP symmetry and T symmetry has the consequence that the rates of mutual transformations of particles are in general different for the direct and inverse reactions (even in stationary states), and are also different when particles are replaced by the appropriate antiparticles. We denote the transition probabilities between the states i and f by ω_{if}, and the probabilities for the CP-conjugate states by $\omega_{\bar{i}\bar{f}}$. We set

$$\omega_{if} = s_{if} + a_{if}$$
$$\omega_{\bar{i}\bar{f}} = s_{if} - a_{if} \tag{3}$$

The CPT invariance implies the following relation: the sum over all final states f vanishes for any initial state i:

$$\sum_f a_{if} = 0 \tag{4}$$

Together with the T symmetry of the probabilities s_{if} and with the equality of particle and antiparticle masses, the condition (4) guarantees the CP symmetry of the equilibrium stationary state ($n_i^0 = \bar{n}_i^0$, $dn_i/dt = 0$). But in a nonstationary state $n_i \neq \bar{n}_i$. Introducing the notation

$$n' = n'^s + n'^a$$

$$\bar{n}' = n'^s - n'^a$$

we obtain the order-of-magnitude estimate (Sec. III)

$$n'^a \sim \frac{a^*}{s} n' \tag{5}$$

where s is the part of the probability of particle transformation which does not depend on replacing particles with antiparticles, and a^* is the antisymmetrical part of that probability [Eq. (3); the asterisk has been used in order not to confuse it with the scale a]. In Eq. (2), $\tau \sim 1/s$, i.e.,

$$n'^a \sim \frac{a^* n^0}{s^2} \Delta t \tag{6}$$

In Sec. IV we obtain for a simplified model of the theory the estimate

$$s \sim \alpha M_c$$

$$a^* \sim \alpha^3 \vartheta^3 \delta_a M_c$$

where $\alpha = g^2$ is the coupling constant of the gauge field, ϑ is a quantity of the order of the Cabibbo angle, and δ_a is the complex phase describing the mixing of quark states.

Making use of $n_c^0 \sim M_c^2$ we obtain the estimate

$$n'^a = \frac{\alpha \vartheta^3 \delta_a M_c^2}{\Delta t} \tag{7}$$

The residual baryon and lepton numbers appear as a result of reactions with the participation of the leptoquark boson [four-boson reactions in Sec. II, Eqs. (13R) and (13R')]. Their probabilities are of the order $\omega \sim \alpha^2 M_c$. Integrating with respect to time [Eq. (15)] we obtain the residual baryon (or lepton) number in the comoving volume $[a(t)]^3$. The factor Δt appears on account of the integration

$$N_B = a^3 n_B \sim \omega n'^a a^3 \Delta t \sim \alpha^3 \vartheta^3 \delta_a M_c^3 a^3 \tag{8}$$

The particle number in the comoving volume has the same order of magnitude as the number of quanta of the background (relic) radiation, i.e., $n^0 a^3 \sim M_c^3 a_c^3$. The baryon asymmetry has the order of magnitude

$$A = \frac{N_B}{N_{\bar{l}}} \sim \alpha^3 \vartheta^3 \delta_a \tag{9}$$

The lepton asymmetry must be of the same order of magnitude. If one assumes conservation for the total number of leptons and quarks (cf. Sec. II) one obtains

$$N_{\bar{l}} - N_l = 3N_B$$

At present there exist no methods which would allow one to verify this relation.

II. VIOLATION OF BARYON AND LEPTON NUMBER CONSERVATION

In unified theories of strong, weak, and electromagnetic interactions one postulates the existence of a so-called leptoquark vector boson, which when emitted or absorbed converts quarks into leptons and vice versa.

Restricting ourselves to theories for which the quarks are postulated to obey exact "color" symmetry and to have fractional electric charges, we have to attribute a fractional charge also to the leptoquark boson. We shall use the following notation: W_c denotes a leptoquark with charge $-1/3$ or any charge differing from this by an integer (if one assumes that the electron is the "particle," these charges are 2/3 and 5/3); $\bar{W}_c$ denotes a leptoquark with charge $+1/3$, or a charge differing from 1/3 by an integer. The bosons W_c and $\bar{W}_c$ interact with the quarks q and leptons l according to the following basic reactions ($\bar{q}$ and $\bar{l}$ denote antiquarks and antileptons):

$$W_c \leftrightarrow q + \bar{l} \tag{10}$$

$$\bar{W}_c \leftrightarrow \bar{q} + l$$

In addition to the reactions (10), in the majority of unified theories there are three more reactions ("vertices"), leading to baryon number nonconservation:

$$W_c \leftrightarrow \bar{q} + \bar{q} \tag{11}$$

$$\bar{W}_c \leftrightarrow q + q$$

the three-boson interaction (an off-mass-shell reaction!):

$$W_c + W_c + W_c \leftrightarrow \text{vacuum} \qquad (12)$$
$$\bar{W}_c + \bar{W}_c + \bar{W}_c \leftrightarrow \text{vacuum}$$

and the four-boson interaction with the participation of three W_c which allows for the on-shell reactions:

$$\bar{W}_c + \bar{W}_c \leftrightarrow W_c + R \qquad (13R)$$
$$\bar{W}_c + R \leftrightarrow W_c + W_c \qquad (13R')$$

Here R is a "regular" vector boson responsible for the weak interactions, such as $W_\pm$ or another gauge boson of zero or integer charge.

Figures 1 and 2 are proton decay diagrams involving the vertex (11) ($p \to \pi + l$) and the vertex (12) ($p \to 3l$). A similar decay for which the vertex (13) is responsible is $p \to 3l + W_+$.

The three-boson vertex (12) was postulated in Ref. 1. In gauge field theories with nonabelian gauge group the three-boson and four-boson interactions (12) and (13) follow from first principles!

Models are possible in which interactions of the type (11) are ab-

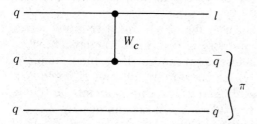

Figure 1

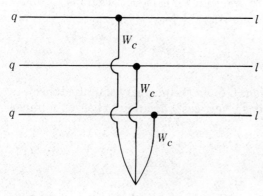

Figure 2

COSMOLOGY

sent, and which lead, in addition, to a strict conservation law of the total number of quarks and leptons (the combined charge N_0):

$$N_0 = N_q + N_l - N_{\bar{q}} - N_{\bar{l}} = \text{const} \tag{14}$$

Such models seem to be preferable. Even for masses M_c which are not too large the proton in such models has a long lifetime—a lifetime which according to experiments (cf. Ref. 5) exceeds 10^{30} years. An estimate (without essential numerical coefficients) for the process depicted in Fig. 1, yields

$$\tau_1^{-1} = \left(\frac{\alpha}{\pi}\right)^2 \left(\frac{M_p}{M_c}\right)^4 M_p$$

[here M_p is the proton mass, M_c is the leptoquark boson (W_c) mass], and for the process depicted in Fig. 2,

$$\tau_2^{-1} = \left(\frac{\alpha}{\pi}\right)^4 \left(\frac{M_p}{M_c}\right)^{12} M_p$$

The first equation requires $M_c > 10^{14} M_p$, and the second $M_c > 3 \times 10^4 M_p$.

We introduce the approximate quantum number

$$r = \tfrac{1}{2}(N_q + N_{\bar{l}} - N_{\bar{q}} - N_l) + N_{W_c} - N_{\overline{W}_c}$$

It is easy to verify that the conservation of γ (and thus of the lepton and baryon numbers), in the absence of the interaction (11), is violated only by the reactions (12) and (13). With these assumptions the following estimate holds for the residual baryon number in a comoving volume $[a(t)]^3$:

$$N_B = \int_0^\infty dt\, a^3 \left[\frac{1}{2}\sum_{ij}(\sigma_{ij}v)(\bar{n}_i\bar{n}_j - n_i n_j) + \sum_i (\sigma_i v)' n_R(\bar{n}_i - n_i)\right] \tag{15}$$

Here n_i and $\bar{n}_i$ are the densities of the W_c and $\overline{W}_c$ of the three different sorts (with charges $\mp\tfrac{1}{3}$, $\pm\tfrac{2}{3}$, $\pm\tfrac{5}{3}$; $(\sigma_{ij}v)$ and $(\sigma_i v)$ are the average values of the products of relative velocities of colliding particles by the cross sections of the reactions (13R) and (13R'); n_R is the density of bosons different from W_c and $\overline{W}_c$. In order to determine n_i it is necessary to solve the kinetic equations.

III. THE KINETIC EQUATIONS

During the early stage of the expansion of the universe the deviations from the equilibrium state are small:

$$n_j^1 = n_j - n_j^0 \ll n_j^0$$

We write the linearized kinetic equations in the following approximate form, for the purpose of obtaining estimates [one can write a more exact system of integral equations for spherically symmetric density functions $n_j^1(p)$ in momentum space]:

$$(S_{ij} + A_{ij})n_j^1 = m_i$$
$$m_i = \frac{1}{a^3} \frac{d}{dt} (a^3 n_i) \tag{16}$$

It is assumed that the matrix S_{ij} does not change under particle-antiparticle conjugation, whereas the matrix A_{ij} changes sign under this substitution. Neglecting in the expression for m_i the deviation of n_i from n_i^0 we obtain for n_i^1 a system of linear algebraic equations with a known right-hand side m_i^0.

We note (although this is essential only for calculations which are more accurate than our estimates) that the matrices A, S, and $A + S$ are singular (have a vanishing determinant), since there is a "singular" direction in the vector space of the n_j^1, for which the right-hand side of (16) becomes zero. This direction corresponds to a variation of the temperature of the equilibrium state

$$\delta n_j^1 = \frac{\partial n_j^0}{\partial T} \delta T$$

We denote the unit vector of this direction by

$$e_j^0 \sim \delta n_j^1$$

We introduce a complete set of orthonormal vectors $e_j^\alpha e_j^\beta = \delta_{\alpha\beta}$ and the corresponding new coordinates $\eta_\alpha = e_j^\alpha n_j^1$. The coefficient η_0 is not determined by Eq. (16), but this is not essential, since e^0 is invariant with respect to particle-antiparticle conjugation. It suffices to find a component of n' orthogonal to e^0.

The matrices

$$\widetilde{S}_{\alpha\beta} + \widetilde{A}_{\alpha\beta} = e_i^\alpha (S_{ij} + A_{ij}) e_j^\beta \qquad \alpha, \beta \neq 0 \tag{17}$$

defined in the space orthogonal to e^0, are nonsingular. Taking into account the fact that $A \ll S$, the inverse matrix becomes

$$(\widetilde{S} + \widetilde{A})^{-1} \approx \widetilde{S}^{-1} + P$$
$$P = -\widetilde{S}^{-1}\widetilde{A}\widetilde{S}^{-1} \tag{18}$$

Setting $n^1 = n^{1S} + n^{1A}$ and $\bar{n}^1 = n^{1S} - n^{1A}$, we obtain

$$n_j^{1A} = e_j^\alpha P_{\alpha\beta} e_k^\beta m_{0k} \tag{19}$$

We estimate the elements of the matrices S, A for a simplified model of the theory.

IV. A MODEL OF THE THEORY

Let n_1, n_2, n_3 denote the densities of the three kinds of leptoquark bosons W_c with the charges $-\frac{1}{3}$, $\frac{2}{3}$, $\frac{5}{3}$, respectively, and let $\bar{n}_1$, $\bar{n}_2$, $\bar{n}_3$ denote the densities of the $\overline{W}_c$. We assume the masses of the different kinds of W_c to be different, with a mass difference of the order of M_c ($\Delta M_c \sim M_c$). The large mass difference between the strange and the charmed quark ($\Delta m_q \sim m_q$) makes this assumption a little more likely by analogy.

The W_c bosons of different kinds undergo transformations via the reactions

$$W_c^i + \bar{q} \to \bar{l} \to W_c^j + \bar{q}$$

$$W_c^i + l \to q \to W_c^j + l$$

Figure 3a represents a typical diagram for this process. The symmetric part of the cross sections is of the order

$$\sigma_{12}' \sim \frac{\alpha^2}{M_c^2}$$

Taking into account that n^0, the density of quarks, leptons (and any other particles) in the critical phase is of the order M_c^3, and the relative velocity of the particles is of the order one, we obtain $S_{12} \sim \alpha^2 M_c$. In making the mechanism for CP violation more concrete we follow the paper of Kobayashi and Maskawa [6]. These authors have found that if one generalizes the Cabibbo mixing model to three or more mixed states, then in the presence of complex mixing matrices there appear CP-violating effects.

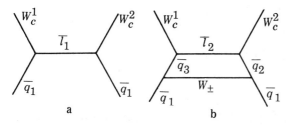

Figure 3

Let us consider mixing among three quark doublets. The states

$$\begin{pmatrix} p_1 & p_2 & p_3 \\ n_1 & n_2 & n_3 \end{pmatrix}$$

are by definition diagonal for the mass operator. The states

$$\begin{pmatrix} P_1 & P_2 & P_3 \\ N_1 & N_2 & N_3 \end{pmatrix}$$

enter into the expressions of the quark and quark-lepton currents. The states P, N and p, n are unitarily related: $P = U_1 p$, $N = U_2 n$. The mixing of the leptons can be described by analogous matrices, but for definiteness we assume that $U_{1l} = U_{2l} = 1$. If one does not consider leptoquark currents, then U_1 and U_2 appear only in the combination $U_1^{-1} U_2$. The asymmetric part of the cross section is due to the interference between the contributions of the diagrams of the type of Fig. 3a and 3b, contributions which differ by the phases $\delta_s + \delta_a$ for particles, $\delta_s - \delta_a$ for antiparticles. The effect is proportional to

$$\cos(\delta_s - \delta_a) - \cos(\delta_s + \delta_a) = 2 \sin \delta_s \sin \delta_a$$

According to the Feynman rules $\delta_s = \pi/2$; the phase δ_a is the parameter of the theory and does not depend on the choice of phases for the quark and lepton states.

An estimate of the asymmetric part yields the order of magnitude

$$\sigma_{12}^a \sim \frac{|M_a| |M_b|}{M_c^2} \sin \delta_a$$

M_a and M_b are the respective contributions of diagrams of the type of Fig. 3a and 3b to the amplitude; $M_a \sim \alpha$, $M_b \sim \alpha^2 \vartheta^3$ (three vertices with change of particle kind). Taking into account the fact that $A_{12} \sim n^0 \sigma_{12}^a$, we obtain

$$A_{12} \sim \alpha^3 \vartheta^3 \delta_a M_c$$

The Kobayashi–Maskawa mechanism is ineffective for extreme relativistic quark energies. Therefore these estimates are valid only if there exist hypothetical quarks with masses of the order of M_c.

The probability of the reactions $W_c \to q + \bar{l}(S_{12}, S_{24}, S_{34})$ is of the order of αM_c. The probability of the four-boson reactions (13R) and (13R') is of the order

$$\omega_R \sim \alpha^2 M_c$$

In the general case, if the masses, the decay probabilities, and the probabilities of the reactions R for the three kinds of W_c do not coincide, then the orders of magnitude of $n^{1a}_{1,2,3}$ are determined by equations (19), (18), and (16):

$$n^{1a}_{1,2,3} \sim Pm$$

$$P \sim \frac{A_{12}}{(s_{14})^2}$$

$$m \sim \frac{n_c}{\Delta t_c}$$

whence

$$n^{1a}_{1,2,3} \sim \alpha \vartheta^3 \delta_a \frac{M_c^2}{\Delta t_c}$$

With the same assumptions, Eq. (15) yields the order of magnitude for N_B,

$$N_B \sim a_c^3 n^{1a}_{1,2,3} \omega_R \Delta t_c \sim a_c^3 n_c^0 \alpha^3 \vartheta^3 \delta_a$$

i.e.,

$$A = \frac{N_B}{N_{\bar{l}}} \sim \alpha^3 \vartheta^3 \delta_a \qquad (20)$$

However, if the difference between the W_c boson masses is smaller than the masses themselves, there appears a new parameter of smallness $C(M_{1c}, M_{2c}, M_{3c})$. The quantity N_B vanishes if the masses of two of the three kinds of leptoquarks are equal. Assume, for instance, that $M_1 = M_2 \neq M_3$. We then automatically have $m_1^0 = m_2^0 \neq m_3^0$ [we recalled that $m_i^0 = a^{-3} d(a^3 n_i^0)/dt$ is the right-hand side of Eq. (16)]. The number of asymmetric transitions from state 1 into state 3 and from state 3 into state 2 are equal to one another. The numbers of symmetric transitions from state 3 into the states 1 and 2 are also equal, and the number of symmetric transitions between the states 1 and 2 is zero. We have

$$n_1^{1s} = n_2^{1s} \neq n_3^{1s}$$
$$n_1^{1a} = -n_2^{1a}$$
$$n_3^{1a} = 0$$

We assume in addition that all the transition probabilities from states 1 and 2 into other states are equal, including the equality of the reactions (13) with change baryon and lepton number. We obtain

$$N_B = 0 \quad \text{for } M_1 = M_2 \neq M_3$$

An example of a function C exhibiting such properties and which is symmetric in its arguments is

$$C = \frac{(M_1 - M_2)^2(M_2 - M_3)^2(M_3 - M_1)^2}{M_1^6 + M_2^6 + M_3^6} \tag{21}$$

We obtain finally

$$A = \alpha^3 \vartheta^3 \delta_a C(M_1, M_2, M_3) \tag{22}$$

where C is a function of the type (21).

V. A MULTIFOLIATED MODEL OF THE UNIVERSE

In 1969 the author included the assumption of neutrality of the universe in the exactly conserved charges, which he took to be the electric charge and the "combined" lepton-baryon charge [of the type of N_0 in Eq. (14) of the present paper], in the proposed cosmological hypothesis of a multifoliated universe (many-sheeted universe) [7]. Another assumption of this hypothesis was a spatial metric that is flat on the average on a large scale, i.e., an infinite curvature radius of the universe. These two assumptions make possible an infinite-fold repetition of cosmological expansion-contraction cycles of the pulsating universe with repetition from cycle to cycle of its statistical characteristics. In any comoving volume the entropy increases in agreement with the second law of thermodynamics, but the increase of entropy from cycle to cycle has no physical meaning and can be removed by a change of scale on the singular hypersurface $t = t_0$, $a(t_0) = 0$. With this redefinition there occurs no change of the densities of exactly conserved charges (electric charge and the combined baryon-lepton charge) or of the integral spatial curvature, since these quantities are assumed to vanish.

The noninvariant charges (the baryon number and the number of leptons in any comoving volume) change, but their ratio to the entropy and the absolute values in the redefined comoving volume are assumed the same at corresponding instants of age of the universe in each cycle.

A dynamical reason for the transition of a flat universe from expansion to contraction could be, in particular, an arbitrarily small (in absolute value) cosmological constant of the appropriate sign ($\varepsilon < 0$, $p = |\varepsilon| > 0$, $\varepsilon + 3p > 0$). In Ref. 7 the formation of black holes was considered as the dynamical mechanism.

We note, among other things, that the assumed repeatability of the statistical characteristics could be an important heuristic requirement which determines the initial inhomogeneities, the densities of entropy

COSMOLOGY 127

and metric, the density distribution of angular momentum, and other statistical parameters of the model.

VI. CONCLUSION

Thus, for a model of the theory we have obtained an estimate of the baryon asymmetry of the universe (the lepton asymmetry is of the same order):

$$A \sim \frac{N_B}{a^3 n^0} \sim C\alpha^3 \vartheta^3 \delta_a$$

Our calculations contain too many uncertainties in order to be able to talk about agreement with experiment, which yields a value of $A \sim 10^{-8} - 10^{-9}$; however, our result is not in contradiction with experiment.

Thus, setting $\alpha = 10^{-2}$, $\vartheta = 0.5$, $\delta_a = 10^{-1}$ and $C = 10^{-2}$ we obtain $A = 10^{-10}$.

The estimate is valid only if there exist hypothetical quarks with masses of the order of M_c. The additional small parameter which appears for the presently known quarks seems to exclude a possibility of agreement with experiment for the concrete model considered in Sec. IV.

The result obtained in the present paper does not depend on the dimensionless parameter $k = 1/M_c G^{1/2}$, which determines the ratio of the duration of the "critical" phase for the process under consideration $\Delta t \sim 1/G^{1/2} M_c^2$, to the characteristic reaction time for the mutual transformation of particles $\tau \sim 1/\alpha M_c$:

$$\frac{\Delta t}{\tau} \sim \alpha k$$

Yoshimura [2] has obtained a formula which differs from ours, according to which the baryon asymmetry A is of the order of k, i.e., is proportional to the duration of the critical phase Δt. This result is in contradiction with the absence of CP violation in a stationary state. As was shown here (Sec. III) small deviations from the equilibrium state, and thus from CP symmetry, are proportional to $1/k$. An integration with respect to time leads to a cancellation of the k dependence. In the paper of Dimopoulos and Susskind [4] it was assumed from the start that $k \sim 1$, and thus the dependence of the result on this parameter is not investigated.

In Sec. V we have argued for the assumption, for the multifoliated model of the universe with statistical characteristics which are repeated every cycle, that the universe is initially neutral.

I am grateful to the participants of the seminar of the Theory Section of FIAN (Lebedev Physics Institute) held on October 13, 1978 for a valuable discussion of a preliminary version of this paper, which I have since corrected. I am particularly grateful to D. A. Kirzhnits and A. D. Linde for reading a preliminary manuscript of this paper and for valuable remarks which led to improvements, and to A. D. Dolgov, who pointed out an error I made in one of the estimates.

Supplement (April 6, 1980)

In 1966 the author advanced the idea of the appearance of baryon asymmetry of the universe during nonequilibrium processes in the cosmological expansion of the "hot" universe as a result of violation of CP invariance, and conjectured violation of conservation of baryon charge. The violation of CP invariance had been discovered long before that, as had the relic radiation that supported the hot model. But the author's hypothesis regarding violation of baryon charge seemed too contrived at the time; this produced a skeptical attitude toward the work as a whole (see, e.g., Zel'dovich and Novikov's book on relativistic cosmology[†]—I do not remember the exact name of the book, and in Gor'kii I have nowhere to look). However, several years before that, in a work then unknown to me, Steven Weinberg substantiated the fundamental possibility of violation of baryon charge because of the absence of a corresponding gauge field. The work may also contain a discussion of the cosmological consequences—I do not know. The author's 1979 paper appeared after Yoshimura's paper of 1978, in which the idea of cosmological development of baryon asymmetry was advanced once more. In this paper the violation of baryon charge is regarded as a consequence of the unified theory of weak, electromagnetic, and strong interactions (the grand unification). In my paper I consider a version of the theory in which there is a $3B + L$ combined law of conservation, but baryon charge is broken by the interaction of gauge fields through diagrams of the type shown in Fig. 4. The SU_5 grand unification scheme, which currently is most popular, allows diagrams of a different type (Fig. 5) with a much higher probability of proton decay.

In my work I have shown that for the theoretically predicted numerical value of the residual baryon asymmetry, the value of the mass

[†]Ya. B. Zel'dovich and I. D. Novikov, *Relyativistskaya astrofizika*, Nauka, Moscow (1967); *Relativistic Astrophysics*, Vol. I, Vol. II, University of Chicago Press, Chicago (1971), trans.—Eds.

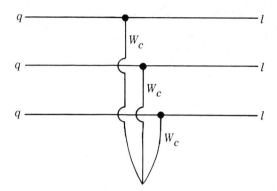

Figure 4

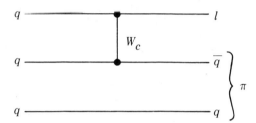

Figure 5

of the quark-lepton boson is of little significance, but it is very important that quarks or Higgs bosons exist with masses of the order of this mass. In this case it is entirely feasible to count on an explanation of the observed asymmetry. In my 1979 paper I unfortunately confined my cosmological discussions to the many-sheeted model of the universe that I had offered previously. In the Comments I briefly discuss other possible models in which the formation of a baryon excess from the neutral hot state is significant, and (in connection with the statistical "reversibility paradox") I also touch on the problem of the "rotation of time's arrow."

REFERENCES

1. A. D. Sakharov, *ZhETF Pis'ma* 5:32 (1967), *JETP Lett.* 5:24 (1967), trans. [Paper 7 in this volume].
2. M. Yoshimura, *Phys. Rev. Lett.* 41:281 (1978).
3. H. Georgi and S. L. Glashow, *Phys. Rev. Lett.* 32:493 (1974).
4. S. Dimopoulos and L. Susskind, SLAC-PUB-2126, June 1978.

5. F. Reines and M. F. Crouch, *Phys. Rev. Lett. 32*:493 (1974).
6. M. Kobayashi and T. Maskawa, *Prog. Theor. Phys. (Kyoto) 49*:652 (1973).
7. A. D. Sakharov, *A Multisheet Cosmological Model*, Preprint of the Institute of Applied Mathematics, Moscow, 1970; [Paper 10 in this volume, trans.].

Paper 12

Cosmological Models of the Universe with Reversal of Time's Arrow

Cosmological models of the universe with reversal of time's arrow are considered. Formulations are given of the hypothesis of cosmological CPT symmetry suggested earlier by the writer, and of the hypothesis of an open model with many sheets, with negative spatial curvature, and with possible violation of CPT symmetry by an invariant combined charge. The statistical paradox of reversibility is discussed for these models. The small dimensionless parameter δ^2/a^2, which characterizes the mean spatial curvature of the universe, is explained as the result of the evolution of the universe through many successive cycles of expansion and contraction.

The equations of motion of classical mechanics and of nonrelativistic quantum mechanics admit time reversal; so also do the equations of quantum field theory (along with the *CP* transformation). The statistical equations, however, are irreversible. This contradiction has been known since the end of the nineteenth century. We shall speak of it as the "global paradox of reversibility" of statistical physics. The traditional explanation ascribes irreversibility to the initial conditions. However, the nonequivalent status of the two directions of time is still retained in the picture of the world.

Present-day cosmology opens up the possibility of eliminating this paradox. The idea of an expanding universe is now generally accepted in cosmology; according to it, a certain instant in time is characterized by the vanishing of the spatial metric tensor (this time of the "Friedmann singularity" will here be denoted for brevity by the symbol Φ). In 1966–1967 the writer suggested that one may consider in cosmology

Source: Космологические модели с поворотом стрелы времени, ЖЭТФ 79:689-693 (1980). Reprinted from *Sov. Phys. JETP* 52:349–351 (1980), translated by W. H. Furry, with permission from American Institute of Physics.

not only later times than Φ, but also earlier times, but then the statistical properties of the state of the universe at the instant Φ are such that the entropy increases not only going forward in time from this instant, but also going backward in time:

$$\frac{dS}{dt} > 0 \quad S(t) > S(0) \quad \text{for } t > 0$$
$$\frac{dS}{dt} < 0 \quad S(t) > S(0) \quad \text{for } t < 0 \tag{1}$$

Thus it is assumed that for $t > 0$ the ordinary statistical equations hold, but for $t < 0$ the time-reversed equations hold. This reversal is valid for all nonequilibrium processes, including those concerned with information, i.e., the processes of life. The author has named this sort of situation the "reversal of time's arrow." Reversal of time's arrow eliminates the reversibility paradox; in the picture of the world as a whole equivalence is restored between the two directions of time, as inherent in the equations of motion.

Despite the absence of dynamical interaction between the regions of the world with $t > 0$ and with $t < 0$, the assumption of the reversal of time's arrow has physical content; some necessary conclusions about the character of the initial conditions at the point Φ follow from it.

As a model example of reversal of time's arrow let us consider the classical kinetic theory of gases. At the time $t = 0$ we postulate a spherically symmetrical velocity distribution of the molecules at each point in space and nonuniform density and temperature distributions in space. We assume (and this is particularly important) that at $t = 0$ there is no correlation between the relative positions and relative velocities of the molecules; in this case this is the "statistical condition" by means of which one proves that the value of the entropy at the point $t = 0$ is a minimum.

In an earlier paper [1] the writer put forward the hypothesis that the universe possesses cosmological *CPT* symmetry. According to this hypothesis, all events in the universe are symmetric relative to the hypersurface that corresponds to the instant Φ of cosmological collapse. Setting $t = 0$ for this instant, we require that there be symmetry under the transformation $t \to -t$. The only exact symmetry that includes time inversion is *CPT* symmetry. It follows from *CPT* symmetry that the point Φ is singular and is neutral with respect to all invariant charges. We shall define *CPT*-conjugate fields on the auxiliary half-space

$$x_0 = |t| \geq 0 \quad -\infty < x_a < +\infty$$

COSMOLOGY

and denote these fields with the indices a and b. We postulate: for spinors, $\psi^a = \gamma_5 \psi^b$; for the components of a unit tetrad, $e^a_{i(j)} = -e^b_{i(j)}$ (*PT* reflection). (The index referred to the tetrad is put in parentheses.)

We map the field a onto the region $t \geq 0$ and the field b onto the region $t \leq 0$ [with the corresponding change of signs of $e^b_{0(j)}$]. From the condition of continuity at the hypersurface we have $e_{\alpha(j)}(0) = 0$ (the point Φ is singular) and $\psi(0) = \gamma_5 \psi(0)$, so that the current vanishes,

$$j(0) = \bar{\psi}\gamma_j\psi = \bar{\psi}\gamma_5\gamma_j\gamma_5\psi = -\bar{\psi}\gamma_j\psi = 0$$

(neutrality condition at the point Φ).

The neutrality of the universe requires that the observed baryon asymmetry arose in the course of nonequilibrium processes of expansion of the universe. For this it is necessary to assume breaking of baryon charge conservation, but it is possible to have conservation of a combined charge of the type $3B \pm L$ (see Refs. 1 and 2), where B is the baryon charge and L is the lepton charge. We note, however, that in the currently most popular schemes for unifying the strong, weak, and electromagnetic interactions [for example, the $SU(5)$ scheme] there is no such conservation (the conservation of $B - L$ is also approximate in most schemes).

CPT symmetry is not the only possible realization of the reversal of time's arrow. It suffices to assume that at the instant Φ the statistical conditions that there be no correlations is satisfied. The most natural assumption is the one according to which violation of *CPT* symmetry in reversal of time's arrow is due to the presence of a finite invariant combined charge (of course, provided such a charge exists and does not possess a gauge field). The numerical size of the combined charge here has no direct connection with the residual baryon asymmetry, which arises dynamically in the course of the expansion of the universe.

The reversal of time's arrow (with or without *CPT* symmetry) is possible either in the ordinary open model of the universe, or in models with infinitely repeating cycles of expansion and contraction (in pulsating models, or, in the present writer's terminology, in many-sheeted models; see Ref. 2). Owing to their inherent singularities, these latter models seem to us more interesting, and we shall consider them in more detail.

First of all we emphasize that in these models cycles close to the instant Φ must be decidedly different from the "later" cycles, for which all the main statistical characteristics asymptotically approach their limiting values for $|n| \to \infty$ (n is the number of the cycle, $-\infty < n < +\infty$). These limiting "self-reproducing" values correspond to the many-sheeted model without reversal of time's arrow (cf. Ref. 2). In the many-sheeted model without reversal of time's arrow, according to Ref. 2, the spatial curvature and all of the invariant charges must be equal to zero (in the sense of average values). In the model with reversal of time's arrow these quantities must become zero only asymptotically. In this sense the many-sheeted sort of model is more general.

Accordingly, let us examine a model with a finite spatial curvature $-a^{-2}$ and, possibly, a finite combined charge. We shall suppose that the curvature is negative (a is the hyperbolic radius), which evidently corresponds to the observations. We shall also assume that the Einstein cosmological constant is different from zero, with its sign corresponding to a vacuum energy density $\varepsilon < 0$. We make no assumption about the absolute value $|\varepsilon|$, but it is very probable that $|\varepsilon|$ is small in comparison with the mean density of matter at the present time. The negative sign corresponds to breaking of the symmetry of the vacuum state with $\varepsilon = 0$.

The dynamics of the universe is determined by the Einstein equation

$$8\pi G T_0^{\,0} = R_0^{\,0} - \tfrac{1}{2} R$$

which we write in the form (with c, the speed of light, set equal to 1)

$$H^2 = \frac{\dot{a}^2}{a^2} = \frac{8\pi G}{3} (\rho + \varepsilon) + \frac{1}{a^2} \qquad (2)$$

where H is the Hubble parameter, ρ is the density of "ordinary" matter, and ρ and $1/a^2$ go to zero as $a \to \infty$. Since $\varepsilon = \text{const} < 0$, at some value of a the quantity H goes to zero and expansion is replaced by contraction. Accordingly, the universe experiences an infinite number of cycles of expansion and contraction.

For the initial conditions in the neighborhood of the point Φ, the following four types of assumption are the most natural (σ is the density of entropy, and n_c is the density of the combined charge; $n_c a^3 = 0$ means that there is no combined charge or that it is equal to zero):

1. $\sigma a^3 \sim 1$ $n_c a^3 \sim 1$
2. $\sigma a^3 \sim 1$ $n_c a^3 = 0$
3. $\sigma a^3 \to 0$ $n_c a^3 \sim 1$
4. $\sigma a^3 \to 0$ $n_c a^3 = 0$

Types 2 and 4 correspond to cosmological *CPT* symmetry. In the case of types 1 and 3, the *CPT* symmetry is broken by the presence of combined charge, which can lead to important differences in the details of the world picture in the positive and negative cycles. Types 1 and 2 correspond to hot models of the universe, types 3 and 4, to cold models. A cold model is the natural realization of the reversal of time's arrow, but on the whole there are neither theoretical nor experimental data for the choice of a definite type.

The entropy σa^3 in a comoving volume a^3 increases in each cycle. Let us suppose that as n increases by 1 the entropy increases by a factor ν; to calculate this number, which is possible in principle, one would have to take into account the main nonequilibrium processes. At present (in "our" cycle n_1) the entropy $\sigma a_1^3 \sim n_\gamma/H^3$, where n_γ is the density of photons of the residual radiation. It is assumed that the density ρ is less than the critical density. For types 1 and 2 we have an estimate of the ordinal number n_1 of our cycle (as an example we have taken $\nu = 1.1$):

$$|n_1| \sim \frac{\ln |n_\gamma H^{-3}|}{\ln \nu} \sim \frac{\ln 10^{87}}{\ln 1.1} \sim 2 \times 10^3$$

In the cold types of model additional cycles are necessary to produce the initial entropy; in type 4 the initial particles arise as the result of a large number of almost empty cycles, owing to the small curvature, proportional to $|\varepsilon|$.

Writing δ^{-3} for the density of the residual-radiation photons, $\delta \sim 0.1$ cm, we have a very small dimensionless number $\delta^2/a^2 \sim 10^{-58}$, which characterizes the curvature of the universe (provided, of course, that the curvature is not identically equal to zero, which still cannot be regarded as excluded). An important advantage of the many-sheeted model with reversal of time's arrow is the possibility of explaining in a natural way the appearance of this dimensionless number in the course of successive cycles of expansion and contraction.

The asymptotic situation with completely similar successive cycles is described by Eq. (2) with the term $1/a^2$ neglected. The solution of Eq. (2) is of the form

$$a = a_n \left(\sin \frac{3}{2} a_0^{-1} t\right)^{2/3}$$

$$a_0 = \left(\frac{8\pi G}{3} |\varepsilon|\right)^{-1/3}$$

The maximal hyperbolic radius a_n of the nth cycle is determined from the condition $\rho(a_n) = |\varepsilon|$, and is proportional to $\nu^{|n|/3} \to \infty$ as $|n| \to \infty$. The duration of each cycle is $T_A = 2\pi a_0/3$. The densities of baryons, of leptons, and of entropy at corresponding times in successive cycles does not depend on $|n|$. The cycles closer to Φ are described by Eq. (2) with ρ neglected (except in relatively small intervals of time at the beginning and end of each cycle). Neglecting ρ, we have $a = a_0 \sin(t/a_0)$, and the duration of each cycle is $T_I = \pi a_0$. The transition from the initial to the asymptotic situation is defined by the condition $\rho(a_0) = |\varepsilon|$, and will occur at cycle number $n_2 > n_1$ (on the assumption that at present $\rho < \rho_c$). The baryon asymmetry n_B/n_γ, however, already has its asymptotic value, since it is determined by the initial stage of the expansion of the universe.

The stability of this pattern of successive collapses has not been investigated. In this paper we have discussed the reversibility paradox, the hypothesis of cosmological CPT symmetry, and the various types of many-sheeted models.

I express my gratitude to all who have taken part in discussing preliminary versions of this paper, and to my wife, E. G. Bonner, for her help.

REFERENCES

1. A. D. Sakharov, *ZhETF Pis'ma* 5:32 (1967); *JETP Lett.* 5:24 (1967), trans. [Paper 7 in this volume].
2. A. D. Sakharov, *ZhETF* 76:1172 (1979); *Sov. Phys. JETP* 49:594 (1979), trans. [Paper 11 in this volume].

Paper 13

Maximum Temperature of Thermal Radiation

American investigators have recently observed cosmic radio emission with effective temperature 3.5 K at a wavelength 7.3 cm [1]. If the assumed thermal character of this radiation (maximum at 0.1 cm) is confirmed, then it will be most natural to regard it as the residual photon field remaining from the initial singular state of the expanding universe, which in this case must be assumed to possess an infinite entropy density ("hot" model of expanding universe; in this model it is necessary to postulate, besides the presence of the photon field, also the presence of a residual graviton gas and of gas made up of pairs of two species of neutrinos with approximately the same average energy and nonthermal spectrum).

In connection with these hypotheses it is of interest to consider the properties of hot matter at very high densities, including such that the gravitational interaction of the photons is significant (the number of photons per unit volume is of the order of the gravitational unit $n_0 = c^{3/2}\hbar^{3/2}G^{-3/2} = 2.4 \times 10^{98}$ cm^{-3}; cf. Ref. 2 by the author, where cold matter is considered).

We denote the energy density by ε. The total energy in a sphere of radius R, separated in isotropic space, contains terms proportional to different powers of R,

$$E = \frac{4\pi}{5} R^3 \dot{R}^2 \varepsilon - \frac{32\pi^2 G R^5}{15}\varepsilon^2 + \frac{4\pi}{3}\varepsilon R^3$$

(we put $c = 1$).

The first two terms are $\sim R^5$ (their sum vanishes for flat space). To

Source: О максимальной температуре теплового излучения, Письма в ЖЭТФ 3:439–441 (1966). Reprinted from *JETP Lett.* 3:288–289 (1966), with permission from American Institute of Physics.

find the value of ε it is necessary to separate the terms proportional to the volume in the gravitational interaction between particles (correlation and exchange interactions).

Neglecting for simplicity all effects of production of baryon and lepton pairs, we have within the framework of gravitational perturbation theory the power expansion

$$\left(\frac{n}{n_0}\right)^{2/3} = G\hbar^{-1} c^{-1} n^{2/3}$$

(we put below $\hbar = c = 1$, $n_0 = G^{-3/2}$) and

$$\varepsilon = An^{4/3} - BGn^2 - CG^2 n^{8/3} \tag{1}$$

The first term is the Stefan–Boltzmann expression; the second is an exchange correlation which decreases the energy for the attracted bosons. The next terms make up the correlation correction, which decreases the energy by virtue of the variational principle (for both bosons and fermions). The coefficients A, B, and C are ~ 1 and >0.

For n of the order n_0 and more, perturbation theory is not valid, but there is no doubt that the energy per photon ε/n cannot have a smaller order of magnitude than the energy of the gravitational interaction of two "neighboring" photons with energy ε/n,

$$\frac{\varepsilon}{n} \gtrsim G \left(\frac{\varepsilon}{n}\right)^2 n^{1/3}$$

i.e.,

$$\varepsilon \lesssim n^{2/3} n_0^{2/3} \tag{2}$$

Thus, at high photon-gas densities the increase of ε like $n^{4/3}$ gives way to a slower growth, like $n^{2/3}$, and the derivative $d\varepsilon/dn$ reaches a maximum at a certain point $n \sim n_0$, after which it decreases [the inequality (2) does not exclude likewise a decrease of the quantity ε itself; this question, which is of importance in cosmology, is more complicated than the question of the derivative $d\varepsilon/dn$].

Neglecting photon interaction we get $n = 0.244 T^3$ and the entropy density is $S = 0.874 T^3$ (the temperature T is in cm^{-1}), i.e., $S = 3.58n$. This proportionality of S to n remains in effect also in the presence of photon interaction, since the total number of photons is an adiabatic invariant of the compression, and consequently the density of the photons and the entropy density are inversely proportional to the volume during adiabatic contraction.

According to thermodynamics, $T = (\partial \varepsilon / \partial S)_V$, i.e. (for thermal radiation), $T = (d\varepsilon/dn)/3.58$ and reaches a maximum T_{max} of the order of the gravitational unit T_0 when n is of the order of n_0:

$$T_0 = k^{-1}c^{5/2}\hbar^{1/2}G^{-1/2} = 1.42 \times 10^{32} \text{ degrees}$$

T_{max}, of the order of T_0, must be regarded as the absolute maximum of the temperature of any substance in equilibrium with radiation.

The foregoing reasoning may, of course, turn out to be inconsistent if it becomes necessary to review the fundamental principles or the fundamental premises of physics at $n \sim n_0$.

REFERENCES

1. A. A. Penzias and R. W. Wilson, *Astrophys. J.* 142:419 (1965).
2. A. D. Sakharov, *JETP* 49.345 (1965), *Sov. Phys. JETP* 22.241 (1966), trans. [Paper 6 in this volume].

Commentary: J. D. Bjorken*
Elementary Particle Physics

While the world at large respects and honors Andrei Sakharov for his vision and courage in dealing with the fundamental problems of contemporary society and of basic human rights, the physics community has great reason to honor and respect his scientific contributions as well. As an elementary particle physicist, my purpose here is to describe to you Sakharov's contributions to that field. These contributions exhibit on the whole an approach which parallels what we see in his work on social issues—a concern with fundamental issues, and an optimistic and often visionary viewpoint, but a viewpoint nevertheless tempered with pragmatic elements as well.

I will not try here to systematically cover all of Sakharov's contributions but rather mention two examples which may help to illustrate the breadth of his contributions and interests and also his approach. The first example has to do with connecting the ideas of CP violation to the baryon asymmetry of the universe. The idea emerged from two experimental discoveries. The first was the discovery in 1964 by Christenson, Cronin, Fitch, and Turlay [1] that the meson known as K_L decays into $\pi^+\pi^-$, implying that the combined symmetry operation of charge-conjugation and parity is not a symmetry of nature. This discovery was recognized right away to be of profound significance. A major implication of the discovery of CP violation is that particle and antiparticle can have different decay rates into their charge-conjugate channels (e.g., $K^+ \to \pi^+ + \pi^+ + \pi^-$ has a very slightly different decay width than does $K^- \to \pi^+ + \pi^- + \pi^-$) and that the conventional weak interaction cannot by itself be the explanation; a new mechanism needs to be invoked. The other important element leading to Sakharov's

---

*Fermi National Accelerator Laboratory, Batavia, Illinois
Source: Talk given at the American Physical Society Special 60th Birthday Festschrift Symposium to Honor Andrei D. Sakharov, January 26, 1981, New York.

arguments was the discovery by Penzias and Wilson [2] in 1965 of the 3° cosmic background radiation. This provided support for the "big-bang" cosmology or, as Sakharov puts it, the "hot" model of the early universe.

In 1966 and 1967, shortly after these results, Sakharov examined the implications of the discovery of CP violation for the prehistory of the hot early universe [3]. An immediate question is very simple: Why is the universe made of matter, i.e., why is there a matter-antimatter asymmetry? In the by-now-standard hot big bang scenario (which in 1967 was not at all standard) this is followed by a more subtle question. When one traces the history of the universe backward to times so early that temperatures are large compared with the proton rest energy, matter and antimatter both are in equilibrium with radiation. Statistical equilibrium then implies a comparable number density, or entropy, for each species. The net amount of matter existing now is determined by the excess amount of matter over antimatter at that time; the calculations show that this excess was in fact only one part in 10^9 or 10^{10}. This peculiar number, the baryon asymmetry, invites an explanation. Sakharov boldly connected this small number with the phenomenon of CP violation. This is not a straightforward affair; quite a few conditions must be met. First of all, Sakharov did not hesitate to use a very early starting time t and temperature T, $t \sim 10^{-43}$ sec, $T \sim 10^{19}$ GeV, when effects of quantum gravity are significant. Next Sakharov assumed that mechanisms responsible for the observed CP violation include a mechanism ultimately allowing for nonconservation of baryons (and leptons). Thus there emerges the possibility that, owing to C asymmetry in microscopic laws, a universe initially symmetric with respect to interchange of particle and antiparticle could evolve asymmetrically into the present situation, where matter apparently dominates antimatter.

This still is not sufficient; one must assume (and this was recognized already by Sakharov) that there be a period in the early expansion where the constituents are out of thermal equilibrium. This could happen if there exist superheavy particles which decay asymmetrically; i.e., because of C violation more baryons than antibaryons are created in the decay processes. Sakharov alludes to M. A. Markov's [4] "maximons," hypothetical particles of the order of the Planck mass, as an example of how this might happen. A second example appeals to a model involving speculations which appear to the contemporary observer somewhat exotic; in 1967 they were in fact not a part of the language at all.

Nevertheless, the model utilizes concepts which pervade the language of the contemporary elementary particle theorist. For example, the argumentation uses the concept of a massive neutrino, whose mass

Sakharov constrained to be ≲30 eV from cosmological considerations. Baryon decay was assumed to be mediated by fractionally charged ("leptoquark") bosons with Yukawa couplings to quarks and leptons. These bosons possessed trilinear self-couplings which allowed an effective six-quark interaction similar to those entertained by Pati and Salam [5] in the early 1970s and by others since. While Sakharov assumed baryon-number violation, he retained conservation of a linear combination of B and L ($3B - L$), similar to the present theoretical tendency for $B - L$ to be conserved. Sakharov's estimate for proton lifetime was $\sim 10^{50}$ years.

After 1967, there was relatively little development of this idea, although it was certainly well known within the U.S.S.R. The notion of neutron-antineutron oscillations did emerge in the U.S.S.R. in the late 1960s from work of V. Kuzmin [6]. But outside the U.S.S.R. this work was little noticed. (I learned of it but did not pay enough attention to it, nor did I communicate it effectively enough to others.) It took the remarkable theoretical advances of the 1970s to really put Sakharov's ideas into motion. What made this possible was the clear emergence of the quarks as basic building blocks of hadrons and of a simplified view of the interactions of quarks at short distance, in particular, the notions of "asymptotic freedom" and quantum chromodynamics. In addition, the gauge theories of electroweak interactions also provide a simplified and quantitative description of the weak force at high energies. With these developments, it is now possible to anticipate the behavior of matter at the extreme temperatures involved in Sakharov's arguments. Indeed the culmination of the progress of the 1970s in providing theories of strong and electroweak interaction lies in the ideas of "grand unification," namely that at energies $\sim 10^{15}$ GeV the strong force is synthesized with the electroweak. The elements which go into this synthesis are in fact quite similar to those required by Sakharov's view of the origin of the baryon-number asymmetry: superheavy bosons of fractional charge coupling quarks to leptons, which induce proton decay and which somehow (although here the situation is even today quite murky) induce CP-violating effects as well.

Thus more than a decade after Sakharov's prophetic contribution, his ideas have become a major research enterprise, involving many people [7]. This includes Sakharov himself [8], who has continued to contribute to the subject within the last few years. He emphasized in particular that existence of a baryon asymmetry implies the necessity of a period in the evolution of the early universe where some of the matter is not in thermal equilibrium. By now the numbers and concepts have been greatly refined, especially with regard to the instability of the proton. (In the most popular grand unification scheme, the proton lifetime is $\sim 10^{31\pm 2}$ years, just within range of experiment.) However the portion of the theory involving CP violation remains remote—less so now than for Sakharov in 1967 but remote nevertheless. But the subject has matured to the point that recently there have emerged serious attempts to connect semiquantitatively the magnitude of the baryon asymmetry to laboratory CP-violation effects, in particular for a prediction [9] for the neutron electric dipole-moment:

$$d_N \gtrsim (2.5 \times 10^{-18}) \left(\frac{n_B}{n_\gamma}\right) e \text{ cm}$$

These calculations are themselves rather visionary, but it is itself remarkable that, right or wrong, we can talk about such things with any degree of credibility at all.

The concepts we have been discussing are commonplace nowadays, but they were not at all commonplace in 1966 or 1967. There was controversy on whether the universe is homogeneous: perhaps there are islands of antimatter in distant parts of the universe and they provide an average matter-antimatter symmetry [10]. A popular cosmological model had a limiting temperature at early times of 150 MeV [11], quite inconsistent with Sakharov's starting temperature of 10^{19} GeV. The notion of baryon decay was, to say the least, uncommon. It is quite remarkable to compare Sakharov's work from that setting and from the contemporary setting: it has aged very, very gracefully.

The second of Sakharov's contributions which I would like to mention has to do with a quite different subject—that of mass relations between different species of mesons and baryons in the quark model. Sakharov's interest in this subject dates back to 1966 when he and Zel'dovich [12] first looked for simple relations between properties of quarks and properties of hadrons. Their simple approach, analogous to what one does in atomic or nuclear physics to get level formulas, is now commonplace. But in 1966 it was considered rather naive, and it took considerable boldness to put it forward seriously. But the idea works, and the method survives. I will leave it to Harry Lipkin, who has worked along these lines in parallel with Sakharov, to describe this in detail; he has a quite fascinating story to tell.

I cannot conclude without expressing some personal feelings. First and foremost is regret that Andrei Sakharov is unable to be in the midst of the scientific community. I have had the privilege, in the course of many trips to the Soviet Union, to meet and to discuss a little physics with him. I remember with pleasure the enthusiasm and radiance present in those meetings—even when the circumstances were most trying. Since the incarceration of Orlov and Sakharov, I no longer visit the U.S.S.R. But I, with him, hope for those better times when we may again meet. I give him my warmest greetings on his 60th birthday.

REFERENCES

1. J. Christenson, J. Cronin, V. Fitch, and R. Turlay, *Phys. Rev. Lett. 13*:138 (1964).
2. A. Penzias and R. Wilson, *Astrophys. J. 142*:419 (1965).
3. A. D. Sakharov, *ZhETF Pis'ma 5*:32 (1967); *Sov. Phys. JETP Lett. 5*:24 (1967), trans. *ZhETF Pis'ma 5*:36 (1967); *Sov. Phys. JETP Lett. 5*:27 (1967), trans. [Papers 8 and 7, respectively, in this volume].
4. M. A. Markov, *ZhETF 51*:878 (1966); *Sov. Phys. JETP 24*:584 (1967), trans.
5. J. Pati and A. Salam, *Phys. Rev. Lett. 31*:661 (1973), *Phys. Rev. D8*:1240 (1973).
6. V. Kuzmin, *ZhETF Pis'ma 9*:335 (1970); *Sov. Phys. JETP Lett. 12*:228 (1970), trans.
7. M. Yoshimura, *Phys. Lett. 41*:281 (1978), *42*:746(E) (1979). A. Yu Ignatiev, N. V. Krasnikov, V. A. Kuzmin, and A. N. Tavkhelidze, *Phys. Lett. 75B*:436 (1978). S. Dimopoulos and L. Susskind, *Phys. Rev. D18*:4500 (1978). D. Toussaint, S. B. Treiman, F. Wilczek, and A. Zee, *Phys. Rev. D19*:1036 (1979). J. Ellis, M. K. Gaillard, and D. V. Nanopoulos, *Phys. Lett. 80B*:360 (1979), *82B*:464(E) (1979). S. Weinberg, *Phys. Rev. Lett. 42*:850 (1979). S. Dimopoulos and L. Susskind, *Phys. Lett. 81B*:416 (1979). M. Yoshimura, *Phys. Lett. 88B*:294 (1979). D. V. Nanopoulos and S. Weinberg, *Phys. Rev. D20*:2484 (1979). S. Barr, G. Segre, and A. Weldon, *Phys. Rev. D20*:2494 (1979).
8. A. D. Sakharov, *ZhETF 76*:1178 (1979); *Sov. Phys. JETP 49*:594 (1979), trans. [Paper 11 in this volume].
9. J. Ellis, M. K. Gaillard, D. Nanopoulos, and S. Rudaz, Preprint LAPP-TH-24, CERN-TH-2967 (October 1980).
10. R. Omnès, *Phys. Rep. 3C*:1 (1972).
11. R. Hagedorn, *Astron. and Astrophys. 5*:184 (1970).
12. A. D. Sakharov and Ya. B. Zel'dovich, *Yad. Fiz. 4*:395 (1966), *Sov. J. Nucl. Phys. 4*:283 (1967), trans. [Paper 19 in this volume].

*Commentary: J. Iliopoulos**

The Baryon Number of the Universe

In traditional cosmological models baryons and antibaryons were created in pairs since the Hamiltonian was assumed to respect baryon number conservation. Any net baryon number should be put in by hand as an initial condition. In the so-called "symmetric" cosmologies it was argued that, within some range of temperatures (~ 1 GeV), a phase transition occurs [1] which results in a spontaneous symmetry breaking and thermal radiation becomes unstable against separation of nucleons from antinucleons. The situation was compared to what happens in a ferromagnet where a domain structure appears. According to this view our presence is due to a local fluctuation.

The trouble with this theory is that there is no evidence for the presence of large amounts of antimatter anywhere in the universe [2]. The rare traces of antinucleons detected in cosmic rays [3] are compatible with the estimated production of antimatter in particle collisions and no large-scale annihilations have been observed. Nevertheless, this was the accepted doctrine for many years, and scientists by and large were unwilling to take what seemed to be a rather radical step. The reason is that, in a symmetric cosmological model, where no net baryon number is put in by hand in the initial conditions, the eventual appearance of baryon excess requires the violation of C and CP invariance as well as baryon number conservation. CP was found to be violated in the celebrated experiment of J. H. Christenson, J. W. Cronin, V. L. Fitch, and R. Turlay [4] in 1964. However, the absolute stability of the proton was considered too sacred to be abandoned.

A. D. Sakharov, in 1966, was the first to remark that, with the discovery of CP violation, half of the conditions for the solution of this old problem were already met. In his paper "Violation of CP invariance,

* Laboratoire de Physique Théorique Ecole Normale Supèrieure, Paris

C asymmetry and baryon asymmetry of the universe" he proposes "an approximate character for the baryon conservation law" and points out, anticipating current research by 12 years, that this is the most plausible mechanism for the observed predominance of matter over antimatter. The three papers in which these ideas are developed are extremely lucid and hardly need any comments, so I shall only point out the similarities and differences with today's most commonly accepted theoretical prejudice.

In his first paper on the subject [Paper 7 in this volume], Sakharov postulates that baryon and lepton numbers are not conserved separately, but he still assumes that a certain linear combination remains invariant. Notice that this is precisely what happens in most present-day grand unified theories [5]. He chooses the combination

$$n_c = 3n_B - n_L \tag{1}$$

because it allows for the transformation of a quark into an antilepton. He assumes a local interaction of the general form

$$\mathcal{L} \sim \bar{q}\gamma^\mu l^c \mathcal{A}_\mu \tag{2}$$

where the $\mathcal{A}_\mu$'s are vector fields with fractional electric charge. In his next paper [Paper 8 in this volume], he associates (2) with the violation of CP invariance by introducing a phase difference between the couplings involving ordinary and strange quarks. With the particular choice of n_c given in (1) the $\mathcal{A}$ bosons carry zero n_c number. Separate baryon- and lepton-number violation is guaranteed by the introduction of trilinear couplings among the $\mathcal{A}$'s. We find essentially all these features in the standard grand unified theories. The only difference is that n_c is taken to be

$$n_c = n_B - n_L \tag{3}$$

and the $\mathcal{A}$'s (the superheavy gauge and Higgs bosons) carry nonzero n_c charges. n_c conservation becomes now a consequence of a particular unbroken symmetry of the Lagrangian. In the six-quark model CP violation is often introduced in the form of phase differences between the couplings of the Higgs bosons to the different quark flavors. On the phenomenological level, Sakharov's original model predicts the main decay mode of the proton to be

$$p \to \mu^+ + 2\bar{\nu}_\mu \tag{4}$$

and the amplitude involves three virtual $\mathcal{A}$ exchanges. So the masses of the $\mathcal{A}$ particles do not have to be extremely large, in contradistinction with most grand unified theories, where a single boson exchange mediates the decay. In this respect, Sakharov's proposal was close in spirit to the low-unification-mass models [6], although the detailed mechanism is different.

Twelve years later, with the advent of gauge theories, Sakharov's proposal on the origin of the baryon number of the universe was rediscovered and formulated in terms of the grand unified theories [7]. Sakharov studied the subject again in his 1978 paper [Paper 11 in this volume], where he correctly identified the crucial role of the expanding universe. Notice that this remark was already contained in his first 1966 paper and its significance is very simple: In a stationary universe, where all interactions are in thermal equilibrium, the particle abundances are given by Boltzmann's law, which involves only the particle masses. But CPT invariance guarantees that baryons and antibaryons have equal masses and, therefore, no net baryon number can be produced. In an expanding universe, however, there is a stage in which the CP- or B-violating interactions drop out of thermal equilibrium. The products of a baryon number-violating process have no chance to reproduce, through collisions, the initial state. A net baryon number will remain.

Today, after extensive theoretical and computational work, the subject, although still far from closed, is understood qualitatively. The most probable scenario seems to indicate that the dominant mechanism for generation of baryons is the decay of superheavy Higgs scalar bosons. Since this is the less well-understood aspect of a gauge theory, no precise quantitative model-independent conclusions can be drawn. It is nevertheless remarkable that even a qualitative agreement between theory and observation can be reached, and it is even more remarkable that the main mechanism was successfully predicted, with no sharp theoretical tools available, but with outstanding physical insight, 15 years ago by Andrei Sakharov.

REFERENCES

1. R. Omnès, *Phys. Rep. 3C*, No. 1 (1972).
2. G. Steigman, 1971 Cargèse Lectures. *Ann. Rev. Astron. Astrophys. 14*: 339 (1976).
3. R. L. Golden et al., *Phys. Rev. Lett. 43*:1196 (1979).
4. J. H. Christenson et al., *Phys. Rev. Lett. 13*:138 (1964).
5. H. Georgi and S. L. Glashow, *Phys. Rev. Lett. 32*:438 (1974). H. Georgi, H. R. Quinn, and S. Weinberg, *Phys. Rev. Lett. 33*:451 (1974).
6. J. C. Pati and A. Salam, *Phys. Rev. Lett. 31*:661 (1973). *Phys. Rev. D8*:1240 (1973).
7. M. Yoshimura, *Phys. Rev. Lett. 41*:28 (1978). S. Dimopoulos and L. Suss-

kind, *Phys. Rev. D18*:4500 (1978). D. Toussaint, S. B. Treiman, F. Wilczek, and A. Zee, *Phys. Rev. D19*:1036 (1979). J. Ellis, M. K. Gaillard, and D. V. Nanopoulos, *Phys. Lett. 80B*:360 (1979). S. Weinberg, *Phys. Rev. Lett. 42*:850 (1979). D. V. Nanopoulos and S. Weinberg, *Phys. Rev. D20*:2484 (1979).

For a review, see D. V. Nanopoulos, Cosmological implications of grand unified theories, Lectures given at the International School of Nuclear Physics, *4th Course: Nuclear Astrophysics*, Erice, March 25 to April 6, 1980, CERN preprint TH-2871.

Commentary: L. Susskind*

Matter-Antimatter Asymmetry

In the normal development of science, an idea emerges when an accumulation of background material makes "the time ripe" for that idea. Usually several people will seemingly independently make the new connections and the general community of scientific peers will be more or less ready to accept them.

Sometimes, however, great leaps of the imagination are made by a single individual in an intellectual environment in which the new ideas are as yet inconceivable. These prescient contributions may be ignored for years or even forgotten. In this note I want to call attention to such a grand leap made in 1966 by Andrei Sakharov. Indeed Sakharov put forward the first good reasons to believe that the baryon number may not be an absolutely conserved quantity.

Sakharov's speculations began with the question of why the observed universe appears to have more matter (protons, neutrons, and electrons) than antimatter (antiprotons, antineutrons, and positrons). To be concrete, why is the baryon (and lepton) number of the universe not zero? According to the laws of relativistic quantum field theory, symmetries exist with respect to the interchange of matter and antimatter. For example, the mass of a particle and its antiparticle must be equal. Similarly, if a particle is unstable so must be its antiparticle, with exactly the same lifetime. If a reaction such as neutron β decay can occur, a similar process in which each particle is replaced by its antiparticle can also occur. This symmetry of nature is one of the deepest consequences of the quantum mechanics and the theory of relativity.

Now, how is it, Sakharov asked, that, although the laws of nature exhibit this matter-antimatter symmetry, the observed universe is so grossly asymmetric, containing matter but almost no antimatter? Be-

---

* Department of Physics, Stanford University, Stanford, California

fore Sakharov's bold speculation, the only answer was that God created the universe with more matter than antimatter and that is all there is to it.

While we cannot say for certain that Sakharov's ideas are correct on this subject, the development of modern gauge theories over the past 13 years has led to a scientific environment in which these ideas appear plausible if not inevitable. Indeed, although Sakharov's ideas were temporarily ignored or forgotten, they were rediscovered as part of the normal scientific development of gauge theories around 1978. In explaining Sakharov's contribution it will be very helpful for me to first review the later rediscovery in the context of grand unified gauge theories (GUTS).

By 1978 a great deal of background material existed which I will briefly review:

1. *The Hot Big Bang* That the universe was born in a superhot, superdense big bang was suggested around 1948 by George Camow. However, it was not until 1965 that Penzias and Wilson verified the Big Bang theory experimentally by observing the remnant cosmic background radiation which was left over from the cosmic explosion. Up to this point, the Hot Big Bang was only one of several competing cosmologies.

2. *The Experimental Discovery of CP Violation* One of the important ingredients in Sakharov's proposal and the later version is the fact that the laws of nature exhibit small asymmetries (CP violation) under matter-antimatter interchange. This was first discovered by Cronin and Fitch in a classic experiment done only one year before Sakharov's paper.

While CP violation is probably a necessary ingredient in producing a net asymmetry, it is not sufficient. Indeed, if the universe began with an equal amount of matter and antimatter, it would remain symmetric unless a mechanism existed which violated the conservation of baryon number.

3. *Nonconservation of Baryon Number* In 1966 there were very few physicists who seriously considered the possibility that conservation of baryon number is violated. The extraordinary stability of the proton and of ordinary matter was generally taken as prima facie evidence for the exact conservation law. This is not to say that a few bold souls did not question the absolute stability of the proton, but no serious reason was offered for a counterproposal. Sakharov was clearly the first to offer a compelling scientific reason for the belief in baryon violation.

Indeed Sakharov's proposal even suggested a rough order of magnitude for the effect.

On the other hand, by 1978 the scientific environment had been radically altered by the extraordinary success of gauge theories of strong and electroweak interactions. Several people, notably Georgi, Glashow, Salam, and Pati, felt that eventually the strong interaction color $SU(3)$ should be unified with electroweak $SU(2) \times U(1)$. Theories of this type were suggested and invariably predicted baryon violation by the exchange of very heavy gauge quanta. For example, in the Georgi–Glashow $SU(5)$ theory there are X bosons with mass $\sim 10^{14}$ GeV which can decay into a quark pair or into a lepton and antiquark. Evidently these particles can mediate baryon-violating processes. However, the rates for such processes at ordinary energies involve the factor $1/M_X^4$, which makes them incredibly small. In any case, by 1978 very serious consideration was being given to the possibility that baryon number violations exist and can be experimentally detected.

4. *The Magnitude of Baryon Excess* At first sight the baryon excess of the universe seems to be an enormously large effect. Indeed it seems almost total! The present universe is almost all matter with at most minute amounts of antimatter. How can such a complete and perfect asymmetry result from the statistical chaos of a hot big bang unless it was put in from the beginning?

To answer this question we must go back to the very early universe when matter was so hot and crowded that hadrons were dissociated into quarks. At such high temperatures the collision of particles will create both quarks and antiquarks. Therefore, we must expect that in the very early hot soup, particles and antiparticles coexisted in abundance. The total number of particles and antiparticles in the soup depends on the temperature having been much higher in the remote past. However, the net baryon number (difference of the number of quarks and antiquarks) is fixed as long as baryon-violating processes are unimportant. What then happened to all the antiparticles as the universe cooled? The answer is that almost all of them found particles with which to combine and annihilate, eventually producing photons. The only particles to survive this annihilation were the excess quarks which could find no mates, and therefore combined into protons and neutrons. In other words, if we want an estimate of the magnitude of the primordial matter excess, we should not compare the present abundance of nucleons and antinucleons but rather the abundance of nucleons with the remnant annihilation radiation. For this reason the discovery of the 3° cosmic background radiation by Penzias and Wilson had an important implication for the magnitude of the primordial excess. We now know that the original excess of quarks was a very small effect since the present ratio of baryons to remanent photons is $Nb/N\gamma \sim 10^{-9}$. Thus in the early universe, for every billion particles there was an excess of about one unit of baryon number.

It was in this environment that Yoshimura in 1978 independently rediscovered Sakharov's proposal that CP and baryon violation would yield a net imbalance. The idea was that at extremely high temperatures baryon-violating processes would occur with significant strength (typically the rate would be of order α at temperatures of order M_X). The violation of CP symmetry should drive these processes in the direction of producing more quarks than antiquarks. Yoshimura attempted an estimate of the effect and found that the imbalance $(Nq - N\bar{q})/(Nq + N\bar{q})$ is most likely small because it involves a rather large number of powers of coupling constants. The empirical result $\sim 10^{-8}-10^{-9}$ did not seem impossible.

Unfortunately, Yoshimura's calculation suffered from an inconsistency. The same calculation also showed that in thermal equilibrium at high temperature an excess of baryons would exist. This, however, was inconsistent with the TCP theorem, which assures that states of positive and negative B are equally weighted in the Boltzmann distribution. Thus, if done correctly, Yoshimura's calculation, based on thermal equilibrium, should have given exactly zero.

Soon after, Dimopoulas and I suggested that in an expanding universe baryon violating processes could be grossly out of equilibrium if they occurred sufficiently early while the expansion was very rapid. The baryon-violating processes would then have to shut down very rapidly with decreasing temperature in order that equilibrium not be reestablished at later times. This requirement could be satisfied if the mass of the X boson is larger than 10^{14} GeV. Baryon violation is essentially shut off and any baryon number produced during the rapid early expansion is frozen in.

A final important element was added by Toussaint, Trieman, Wilczek, and Zee, who suggested a detailed process for the development of baryon excess. These authors suggested that for temperatures greater than M_X the X bosons would exist in thermal equilibrium, with an abundance comparable to ordinary particles such as photons. As the temperature goes below M_X the number of X bosons is exponentially suppressed by the Boltzmann factor $\exp(-M_X/kT)$. The previously abundant X's decay, and since these decays violate baryon conservation and CP, they can preferentially produce quarks relative to antiquarks.

These are the main ideas in our current thinking about baryon

COSMOLOGY   155

excess. What is most remarkable about Sakharov's 1966 paper is that it completely and correctly stated all the essential points I have discussed. In addition to the need for B and CP violation Sakharov correctly understood the role of the CPT theorem and the need for rapid expansion during the baryon-producing era. He also suggested superheavy particles called maximons which play the same role as the X boson of the modern version.

In conclusion I would like to reemphasize the boldness and essential moderness of Sakharov's astonishing 1966 proposal and to state again that this work was the first to give cogent reasons for the belief that the proton might decay.

Part 3

Field Theory and Elementary Particles

Commentary: A. D. Sakharov

Field Theory and Elementary Particles

1. "Vacuum quantum fluctuations in curved space and the theory of gravitation" [Paper 14]. The hypothesis that the zero-point Lagrangian of the gravitational field is a vacuum correction is formulated. The action introduced by Einstein and Hilbert for curved space,

$$S_g = -\frac{1}{16\pi G}\int \sqrt{-g}\, R\, dx$$

is interpreted as the change in the action of quantum vacuum fluctuations accompanying the curving of space (the change proportional to the first power of the components of the curvature tensor). In the paper, it is assumed that the change in the action can be expressed by a quadratically diverging integral over the momenta of the virtual particles. It is assumed that the integral is cut off by unknown physical processes at momenta of order $k_0 \sim 10^{19}$ GeV/c, so that the gravitational constant has the correct order of magnitude

$$G \sim \frac{1}{\int^{k_0} k\, dk} \sim \frac{1}{k_0^2}$$

The curvature-independent part of the Lagrange function (or energy-momentum tensor) of the vacuum corresponds to the cosmological constant, the linear term corresponds to Einstein's theory of gravitation, and the following terms in the expansion describe corrections to Einstein's theory that are nonlinear in R.

A similar idea was formulated earlier by L. E. Parker, who conjectured that the source of gravitational action is a nonlinear nonrenormalizable interaction of vacuum fields [terms of the form $(\bar{\psi}\psi)^2$]. Later these ideas were developed by H. Terezawa.

The interpretation of the cosmological constant as the Lagrange function of elementary vacuum quantum fields was proposed by Zel'dovich. The hypothesis of a zero-point Lagrangian for the electromagnetic field was put forward earlier by Landau and Pomeranchuk and, independently, by Fradkin; for the W boson it was considered by Steven Weinberg, and subsequently the hypothesis of the zero-point Lagrangian for the electromagnetic field and W boson was also considered by Zel'dovich. It is natural to apply the zero-point Lagrangian hypothesis to all boson fields of geometrical origin (g_{ik} and gauge fields, which occur in the expression for the connection). It has also been suggested that the concept may be important for extending the symmetry group in supersymmetry theories.

I regard the assumption of a high degree of divergence (quadratic in the expression for G^{-1} and up to the fourth power of the momentum for the cosmological constant) as a shortcoming of this first paper on the zero-point Lagrangian of the gravitational field.

2. A preprint of the Institute of Applied Mathematics, Moscow (1967) [Paper 15].

3. Seminar at the P. N. Lebedev Physics Institute (1970).

4. "Spectral density of eigenvalues of the wave equation and vacuum polarization" [Paper 16].

The three investigations 2, 3, and 4 develop the theory of Paper 14. The dependence of the energy-momentum tensor and the density of the vacuum Lagrange function on the curvature of space and given external fields is calculated without allowance for the interaction of the particles with one another in the vacuum state (in diagram language, the calculation is made in the single-loop approximation). It is shown that the divergence of the effective Lagrangian in each order in the expansion in powers of the curvature is not greater than logarithmic (for the curvature-independent term, i.e., for the cosmological constant, this was noted independently by Zel'dovich). The expression for the gravitational constant takes the form

$$G^{-1} = (2\pi)^{-1} \Sigma c_i m_i^2 \ln \frac{\Lambda}{m_i^2}$$

where c_i are constants ~ 1, and Λ is the cutoff parameter.

The most perspicuous calculation of the vacuum energy of the scalar field is 3. A hypersphere with time-dependent radius was considered. The difference between the density of the levels of the Helmholtz equation on the hypersphere and the density of the levels in flat space of equal volume leads to a term in the energy proportional to $(1/a^2)\int k\,dk$. The conclusion drawn is that the volume of the hypersphere is $V = 2\pi^2 a^3$. The eigenvalues of the Helmholtz equation $\nabla^2\phi + k^2\phi = 0$ are

$$k^2 = \frac{1}{a^2} J(J+2)$$

$$g = (J+1)^2$$

so that the level density is

$$\frac{dn}{dk} = \frac{dn/dJ}{dk/dJ} = a^3 k^2 \sqrt{1 + \frac{1}{a^2 k^2}}$$

Calculating the correction to the energy $\varepsilon = \int k(dn/dk)\,dk$, we find $\Delta\varepsilon = (V/4\pi^2 a^2)\int k\,dk$. The change in the energy of the field oscillators associated with the nonstationarity can be calculated by means of the self-similar solution for a harmonic oscillator with variable parameters (see comment 1 in Part 2) and is proportional to $(\dot{a}^2/a^2)\int k\,dk$. The sum of the first two terms is proportional to

$$\frac{1}{a^2} + \frac{\dot{a}^2}{a^2} = \tfrac{1}{3}(R_0^0 - \tfrac{1}{2}R)$$

as follows from the requirement of agreement with Einstein's theory. However, the divergent integral must be regularized by separating the terms which depend on the mass of the particles; these have the opposite sign and are proportional to $-m^2 \int dk/k$. Thus, the regularization decreases the degree of the divergence and restores the correct sign of the reciprocal gravitational constant.

In 3 and 4 I developed a Lagrangian formulation based on a generalization of the concept of the density of eigenvalues to an indefinite operator of the d'Alembert type. The eigenvalue density $\rho(\lambda)$ for a definite operator is defined by the requirement

$$\frac{1}{\varepsilon^n}\Big|\sum_0^\infty F_n(\lambda_i \varepsilon) - \int_0^\infty \rho(\lambda) F_n(\lambda\varepsilon)\,d\lambda\Big| \to 0 \quad \text{as } \varepsilon \to 0$$

$$n = 0, 1, 2, \ldots$$

where the weighting function $F(x)$ tends to zero sufficiently rapidly as $x \to 0$ and $x \to \infty$. In the case of indefinite operators, the sum is regularized by a cutoff function which depends on the four-dimensional euclidean radius (with passage to the limit after summation), and the weighting function satisfies the integral relations

$$\int_{-\infty}^{+\infty} F(\lambda)\, d\lambda = 0$$

$$\int_{-\infty}^{+\infty} \lambda F(\lambda)\, d\lambda = 0$$

(for real metric $+---$), as a consequence of which the "density function" $\rho(\lambda)$ is determined up to a linear function of λ. This procedure led to the expressions of the Fock–Schwinger proper-time method derived in a new way. (See also the remarks at the end of Paper 16.)

5. "Scalar-tensor theory of gravitation" [Paper 17]. In the Brans–Dicke scalar-tensor theory of gravitation, the gravitational field is characterized not only by the metric tensor but also a scalar metric scale field $\mu(x)$. It is postulated that the masses of all particles depend on the coordinates and are proportional to $\mu(x)$. It is pointed out in my paper that the relation $G \sim 1/\mu^2$, which follows from the theory of the zero-point Lagrangian, leads to a "degenerate" form of the scalar-tensor theory, all of its consequences being identical to the purely tensor theory of Einstein. As is pointed out by Brans and Dicke, it is only in the case of this degenerate form that the equivalence principle is satisfied for the gravitational mass defect. In the paper, these propositions are illustrated by simple examples.

6. "Topological structure of elementary charges and CPT symmetry" [Paper 18]. This appeared in the 1972 volume dedicated to the memory of I. E. Tamm. At the present time, I am doubtful about the ideas put forward in this paper.

7. "The quark structure and masses of strongly interacting particles" [Paper 19]. The formula is based on a naive model of nonrelativistic quarks, and is constructed in the same manner for mesons and baryons—additively with respect to the masses. The splitting of masses of the same composition is described by the spin-spin interaction of the quarks, which for the strange quark is assumed to be ξ times weaker, where ξ is an empirical coefficient. In 1965–1966, the majority of attempts to find relations between the hadron masses were based on SU_3 or SU_6 symmetry. The Wigner–Eckart group-theoretical formula was used, being applied to the squares of the masses for the mesons and to the masses for the baryons. A confirmation of this approach was seen in relations of the type $\rho^2 - \pi^2 = (K^*)^2 - K^2$ for mesons and the Gell-Mann–Okubo formula for baryons. In my paper with Zel'dovich, the relation for the mesons is satisfied because ξ is approximately inversely proportional to the quark mass. The mass differences of the strange and ordinary quarks $s - o$ and the coefficients ξ determined from the

baryon and meson masses were found to be approximately equal:

$$s - o = \begin{cases} \frac{1}{4}(K + 3K^*) - \frac{1}{4}(\pi + 3\rho) = 194 \text{ MeV} \\ \Lambda - N = 176 \text{ MeV} \end{cases}$$

$$\xi = \begin{cases} \dfrac{K^* - K}{\rho - \pi} = 0.645 \\ 1 - \dfrac{3}{2} \dfrac{\varepsilon - \Lambda}{\Delta - N} = 0.61 \end{cases}$$

The spin-spin interaction which occurs in the mass formula is determined by scalar products $(\sigma_i \sigma_j)$ of spin matrices, which are found from the obvious relation

$$(\sigma_1 + \sigma_2 + \sigma_3)^2 = J(J + 1)$$

(J is the total spin).

The paper also suggests that baryons are constructed from four quarks and one antiquark, a suggestion which is no longer of interest. In particular, it contradicts experiments with high-energy neutrinos and antineutrinos, which prove that the relative weight of the antiquark "sea" is small.

8. "Mass formula for mesons and baryons with allowance for charm" [Paper 20]. The ideas of the previous paper are applied to the description and prediction of the masses of hadrons containing the charmed quark.

To find the eigenvalues of the operator (an 8×8 matrix) of the spin-spin interaction in a hadron containing three essentially different quarks ("ordinary", s, c),

$$H_{\sigma\sigma} = b[\xi_c(\sigma_s \sigma_o) + \xi_c(\sigma_c \sigma_o) + \xi_s \xi_c(\sigma_s \sigma_c)]$$
$$= A(\sigma_1 \sigma_2) + B(\sigma_2 \sigma_3) + C(\sigma_3 \sigma_1)$$

the following device is used. The fourfold degenerate values for spin $\frac{3}{2}$ are found from the condition $(\sigma_i \sigma_j) = \frac{1}{4}$ and are equal to $h_{3/2} = \frac{1}{4}(A + B + C)$. The twofold degenerate values for spin $\frac{1}{2}$ are found from a two-dimensional secular equation which can be readily derived by noting that the three operators $(\sigma_1 \sigma_2)$, $(\sigma_2 \sigma_3)$, and $(\sigma_3 \sigma_1)$ are obtained from each other by rotation through 120° in the two-dimensional plane. It is found that

$$h_{1/2} = -\tfrac{1}{4}(A + B + C) \pm \frac{1}{2^{3/2}} [(A - B)^2 + (B - C)^2 + (C - A)^2]^{1/2}$$

The $\Sigma - \Lambda$ and $\Sigma_c - \Lambda_c$ mass splittings are also described by this formula for $B = C$. The paper predicts that the $\Sigma_c - \Lambda_c$ mass difference (composition u, d, c) is 161.5 MeV, while experimentally the value is 168 MeV (determined from the decay $\Sigma_c^{++} \to \Lambda_c^+ + \pi^+$). The prediction for the difference $D^* - D$ is also fulfilled satisfactorily, but the absolute values of the masses are overestimated.

9. "Mass formula for mesons and baryons" [Paper 21]. The number of parameters in the linear mass formula is reduced. It is assumed that (1) $\xi_c = m_o/m_c$ and $\xi_s = m_o/m_s$, and (2) the coefficient in $H_{\sigma\sigma}$ for baryons can be expressed in terms of the coefficient for mesons: $b_B = b_M/3$. Experimentally,

$$b_M = \rho - \pi = 635$$

$$b_B = 2\,\frac{\Delta - N}{3} = 195.3$$

The coefficient $\frac{1}{3}$ is justified as follows. The spin-spin interaction of the quarks is proportional to $|\psi(0)|^2(g_1g_2)/m_1m_2 \sim (g_1g_2)/V$. Here, V is the effective volume of the hadron, which can be assumed to be 1.5 times greater for a baryon than for a meson in accordance with the total number of quarks in a hadron; (g_1g_2) is the scalar product of the charge vectors in the space of the color charges, and is equal to $-g^2$ for a meson and $-r^{-1}g^2$ for quarks of different colors in a baryon. Here, $r = n - 1$ is the rank of the color group SU_n^c, and for the usually adopted group SU_3^c we have $r = 2$, i.e., $b_B/b_M = V_M/rV_B = \frac{1}{3}$.

10. "An estimate of the coupling constant between quarks and the gluon field" [Paper 22]. Comparison of the electromagnetic splittings in hadrons of the same composition, which are proportional to e_1e_2/m_1m_2, and the gluon splittings, which are proportional to g_1g_2/m_1m_2, makes it possible to estimate the coupling constant g^2 and verify the presence of the factor r^{-1} in the scalar product of the charge vectors for baryons.

11. "Generation of mesons" (when this was published in *Zhurnal Eksperimental'noĭ i Teoreticheskoĭ Fiziki* in 1947, the editorial board changed the title to "Generation of the hard component of cosmic rays") [Paper 23]. Most of the paper is obsolete since it refers to a particular obsolete model of the theory of elementary particles. The paper contains kinematic expressions for the generation thresholds. It is shown that in the case of generation on nucleons in nuclei the thresholds are lowered, and the probabilities of the processes in the additional region are estimated. Generation with the absorption of two photons of the incident

radiation is discussed. In this connection, the scattering of a strong electromagnetic wave by a free electron with doubling of the frequency is discussed. A circularly polarized wave of frequency ω is considered. For $\hbar\omega \ll mc^2$ a classical treatment is admissible. The electron executes a circular motion with the same frequency and emits electromagnetic waves with frequencies ω, 2ω, 3ω, ..., whose intensities are found by expanding the radiated field in a Fourier series. These arguments belong essentially to nonlinear optics, which has been developed in recent years.

12. Candidate's dissertation (P. N. Lebedev Physics Institute, 1947): "Theory of nuclear transitions of the type $O \to O$." Transitions in RaC' and O^{16} are considered. Considering the work now, the following points appear to me to be of interest:

1. The attempt to introduce a new approximate quantum number of charge parity associated with the change in the wave function under the proton ↔ neutron transformation in systems with equal numbers of these particles.

2. A part of the thesis published as the article "Interaction of the electron and the positron in pair production" in 1948 in *Zhurnal Eksperimental'noĭ i Teoreticheskoĭ Fiziki* [Paper 24]. The matrix element is proportional to an integral over the momentum space with weight 1 and is therefore proportional to ψ(0), where ψ is the wave function for the relative distance of the electron and the positron. The additional factor that takes into account the interaction is a function of the relative velocity v in the final state:

$$v \approx \frac{2^{1/2}}{m} (E_+ E_- - p_+ p_- \cos\vartheta - m^2)^{1/2}$$

$$f(v) = \frac{|\psi(0)|^2}{|\psi(\infty)|^2} = \frac{2\pi e^2}{v}\left[1 - \exp\left(-\frac{2\pi e^2}{v}\right)\right]^{-1}$$

(the well-known solution of the nonrelativistic Coulomb problem in parabolic coordinates is used). In the case of pair production accompanying $O \to O$ transitions the differential probability tends to infinity as $v \to 0$.

Paper 14

Vacuum Quantum Fluctuations in Curved Space and the Theory of Gravitation

In Einstein's theory of gravitation one postulates that the action of space-time depends on the curvature (R is the invariant of the Ricci tensor):

$$S(R) = -\frac{1}{16\pi G} \int dx \sqrt{-g}\, R \tag{1}$$

The presence of the action (1) leads to a "metrical elasticity" of space, i.e., to generalized forces which oppose the curving of space.

Here we consider the hypothesis which identifies the action (1) with the change in the action of quantum fluctuations of the vacuum if space is curved. Thus, we consider the metrical elasticity of space as a sort of level displacement effect (cf. also Ref. 1).[†]

In present-day quantum field theory it is assumed that the energy-momentum tensor of the quantum fluctuations of the vacuum $T_k^i(0)$ and the corresponding action $S(0)$, formally proportional to a divergent integral of the fourth power over the momenta of the virtual particles of the form $\int k^3\, dk$, are actually equal to zero.

Recently Ya. B. Zel'dovich [3] suggested that gravitational interactions could lead to a "small" disturbance of this equilibrium and thus to a finite value of Einstein's cosmological constant, in agreement with the recent interpretation of the astrophysical data. Here we are interested in the dependence of the action of the quantum fluctuations on the

Source: Вакуумные квантные флюктуации в искривленном пространстве и теория гравитации, Доклады АН СССР 177:70-71 (1967). Reprinted from *Sov. Phys. Dokl.* 12:1040–1041 (1968), with permission from American Institute of Physics.
[†] Here the molecular attraction of condensed bodies is calculated as the result of changes in the spectrum of electromagnetic fluctuations. As was pointed out by the author, the particular case of the attraction of metallic bodies was studied earlier by Casimir [2].

curvature of space. Expanding the density of the Lagrange function in a series in powers of the curvature, we have (A and $B \sim 1$)

$$\mathscr{L}(R) = \mathscr{L}(0) + A \int k\, dk\, R + B \int \frac{dk}{k} R^2 + \cdots \qquad (2)$$

The first term corresponds to Einstein's cosmological constant.

The second term, according to our hypothesis, corresponds to the action (1), i.e.,

$$G = -\frac{1}{16\pi A \int k\, dk} \qquad A \sim 1 \qquad (3)$$

The third term in the expansion, written here in a provisional form, leads to corrections, nonlinear in R, to Einstein's equations.[†]

The divergent integrals over the momenta of the virtual particles in (2) and (3) are constructed from dimensional considerations. Knowing the numerical value of the gravitational constant G, we find that the effective integration limit in (3) is

$$k_0 \sim 10^{28} \text{ eV} \sim 10^{33} \text{ cm}^{-1}$$

In a gravitational system of units, $G = \hbar = c = 1$. In this case $k_0 \sim 1$. According to the suggestion of M. A. Markov, the quantity k_0 determines the mass of the heaviest particles existing in nature, and which he calls "maximons." It is natural to suppose also that the quantity k_0 determines the limit of applicability of present-day notions of space and causality.

Consideration of the density of the vacuum Lagrange function in a simplified "model" of the theory for noninteracting free fields with particles $M \sim k_0$ shows that for fixed ratios of the masses of real particles and "ghost" particles (i.e., hypothetical particles which give an opposite contribution from that of the real particles to the R-dependent action), a finite change of action arises that is proportional to $M^2 R$ and which we identify with R/G. Thus, the magnitude of the gravitational interaction is determined by the masses and equations of motion of free particles, and also, probably, by the "momentum cutoff."

This approach to the theory of gravitation is analogous to the discussion of quantum electrodynamics in Refs. 4 to 6, where the possibility is mentioned of neglecting the Lagrangian of the free electromag-

---

[†] A more accurate form of this term is $\int (dk/k)(BR^2 + CR^{ik}R_{ik} + DR^{iklm}R_{iklm} + ER^{iklm}R_{iklm})$ where $A, B, C, D, E \sim 1$. According to Refs. 4 to 7, $\int dk/k \sim 137$, so that the third term is important for $R \gtrsim 1/137$ (in gravitational units), i.e., in the neighborhood of the singular point in Friedman's model of the universe.

netic field for the calculation of the renormalization of the elementary electric charge. In the paper of L. D. Landau and I. Ya. Pomeranchuk the magnitude of the elementary charge is expressed in terms of the masses of the particles and the momentum cutoff. For a further development of these ideas see Ref. 7, in which the possibility is established of formulating the equations of quantum electrodynamics without the "bare" Lagrangian of the free electromagnetic field.

The author expresses his gratitude to Ya. B. Zel'dovich for the discussion which acted as a spur for the present paper, for acquainting him with Refs. 3 and 7 before their publication, and for helpful advice.

REFERENCES

1. E. M. Lifshits, *ZhETF* 29:94 (1954); *Sov. Phys. JETP* 2:73 (1954), trans.
2. H. B. G. Casimir, *Proc. Nederl. Akad. Wetensch.* 51:793 (1948).
3. Ya. B. Zel'dovich, *ZhETF Pis'ma* 6:922 (1967); *JETP Lett.* 6:345 (1967), trans.
4. E. S. Fradkin, *Dokl. Akad. Nauk SSSR* 98:47 (1954).
5. E. S. Fradkin, *Dokl. Akad. Nauk SSSR* 100:897 (1955).
6. L. D. Landau and I. Ya. Pomeranchuk, *Dokl. Akad. Nauk SSSR* 102:489 (1955), trans. in Landau's Collected Papers (D. ter Haar, ed.), Pergamon Press, 1965.
7. Ya. B. Zel'dovich, *ZhETF Pis'ma* 6:1233 (1967).

Paper 15

Vacuum Quantum Fluctuations in Curved Space and the Theory of Gravitation

In Einstein's theory of gravitation the dependence of the action of space-time on curvature is postulated as

$$S(R) = -\frac{1}{16\pi G} \int dx \sqrt{-g}\, R \qquad (1)$$

Such an action leads to a "metrical elasticity" of space, i.e., to the appearance of a generalized force which acts against the curving of space.

We shall investigate the hypothesis that the action (1) is identified with the one following from the change of the quantum fluctuations of a vacuum due to the curving of space. Thus we treat the metrical elasticity of space as a kind of level displacement effect (compare also Ref. 1).

In current quantum field theory the energy-momentum tensor of the vacuum quantum fluctuations $T_k^i(0)$ and the corresponding action $S(0)$ (formally proportional to an integral over the momenta of the virtual particles which diverges as the fourth power of momentum $\int k^3\, dk$) are considered to be actually zero.

Recently Ya. B. Zel'dovich [2] assumed that the gravitational interaction can lead to some "small" deflection from these zero values, and hence to some finite value of Einstein's cosmological constant, thus verifying for the recent reinterpretation of astrophysical observations. We are interested here in the dependence of the quantum fluctuations action on the space curvature. Developing the Lagrange function density $\mathcal{L}(R)$ into a power series in the curvature, we have

Source: Вакуумные квантовые флюктуации в искривленном пространстве в теория гравитации, Препринт Института прикладной математики АН СССР, 1967. Preprint, Institute of Applied Mathematics, Moscow, 1967.

$$\mathscr{L}(R) = \mathscr{L}(0) + A \int k\, dk\, R + B \int \frac{dk}{k} R^2 + \cdots \tag{2}$$

The first term corresponds to Einstein's cosmological constant. The next term corresponds, according to our hypothesis, to the action (1), i.e.,

$$G = -\frac{1}{16\pi A \int k\, dk} \qquad A \sim 1 \tag{3}$$

The third term is not essential for our treatment and is written down symbolically.[†]

The divergent integrals over the momenta of the virtual particles in Eqs. (2) and (3) are constructed according to dimensionality arguments. Taking into account the value of the gravitational constant G, we find that the limit of integration in Eq. (3) is effectively

$$k_0 \sim 10^{28}\ \text{eV} \sim 10^{33}\ \text{cm}^{-1}$$

In the gravitational system of units, $G = \hbar = c = 1$. Thus $k_0 = 1$. According to M. A. Markov, the value of k_0 determines the mass of the heaviest possible particle, which he calls a "maximon." It is but natural to assume that the same quantity k_0 also defines the range of validity of the current conceptions of space and causality.

In general relativity a Riemannian geometry of space-time is assumed, with particles moving along geodesics and Eq. (1) valid for the curved space action. By the hypothesis of this paper, the second postulate follows from the first one.

To illustrate the proposed concept of the nature of gravitation, we shall calculate the variation of the action due to the transition from a flat space to a curved one for a "model" theory in which two kinds of fields are present (both fields are free; there is no interaction): c_0, the field of "real" particles with mass M_0 ($c_0 M_0^2 > 0$) and c_1, the field of "ghost" particles with mass M_1 ($c_1 M_1^2 < 0$). M_0 and $|M_1| \sim k_0$; for simplicity we will consider only boson fields (neutral scalar fields) $S(R) - S(0)$ depending on the curvature R and proportional to RM_0^2.

Action is calculated by integration over the Feynman trajectories. By a trajectory for the field ϕ we mean an arbitrary function of space and time. It is convenient to develop this function into a series over the eigenfunction of the wave equation:

[†] More precisely we have $\int (dk/k)(BR^l + CR^{ik}R_{ik} + DR^{iklm}R_{iklm} + ER^{iklm}R_{iklm})$, where A, B, C, D, E ~ 1. According to Refs. 4 and 5, $\int (dk/k) \sim 137$ [compare (A4) in the Appendix]. This term is essential near the Friedmann singularity if $R \geq 1/\int (dk/k) \sim 1/137$ (in G units).

FIELD THEORY AND ELEMENTARY PARTICLES

$$\phi(\xi,t) = \sum \phi_i(\xi,t) z_i \qquad (4)$$

$$\Box \phi_i + \lambda_i \phi_i = 0 \qquad \int_V \phi_i^2 \, dx = 1$$

Here λ_i are the eigenvalues of the wave equation.

The classical action for the "trajectory," defined by a set of numbers z_i, is

$$S = \tfrac{1}{2} \sum_i (\lambda_i - M_b^2) z_i^2 = \sum S_i \qquad (5)$$

From Eq. (5) and the stationarity of the action, $\partial S/\partial z_i = 0$ follows the classical equation of motion,

$$z_i = 0 \qquad \text{for } \lambda_i \neq M_b^2$$

Now we evaluate the Feynman integral over paths,

$$e^{iS_b} = \frac{1}{N} \prod_i \int_{-\infty}^{+\infty} dz_i \, e^{iS_i} \qquad (6)$$

from which the phase S_b (the action of quantum fluctuations of the boson field) is

$$S_b = \frac{\pi}{4} \sum_i \text{sign}\,(\lambda_i - M_b^2) \qquad (7)$$

The analogous formula for fermions differs from Eq. (7) with respect to sign. This corresponds to the negative energy of the "Dirac background":

$$S_f = -\frac{\pi}{4} \sum_i \text{sign}\,(\lambda_i - M_f^2) \qquad (8)$$

The sums S_b and S_f suffer some change when the space is curved. This change is most easily calculated for the simplest case of the curved space-time, that of a space hypersphere of constant radius a. The trace of the Ricci tensor for this case is

$$R = -R_\xi = -\frac{6}{a^2}$$

(R_ξ is the trace of the space tensor; the signature $+---$ is assumed.)

Let us consider a four-volume $V = V_\xi T$ with $0 < t < T$ and $V = 2\pi^2 a^3$.

Substituting $\phi_i = e^{i\omega_n t} \psi_m(\xi)$, we have

$$\Delta \psi_m + k^2 \psi_m = 0$$

$$\lambda = \omega^2 - k^2$$

$$k^2 = \frac{1}{a^2} J(J+2) \quad (9)$$

$$g(J) = (J+1)^2$$

Here $J = 0, 1, 2, \ldots$ is the hyperspherical angular momentum and $g(J)$ is the statistical weight.

Using Eqs. (9) we find the spectral density for the number of eigenvalues λ_i for the volume $dk \, d\omega$:

$$dN_b = dk \, d\omega \, \frac{V}{2\pi} \left(\frac{k^2}{2\pi^2} + \frac{1}{4\pi^2 a^2} \right) \quad (10)$$

Eq. (10) is easily generalized to the case of an arbitrary time-independent geometry by substituting $\frac{1}{6} \int d\xi \sqrt{g(\xi)} R_\xi$ instead of V_ξ/a^2. In the time-dependent case $a(t)$ the spectral density must be taken per volume $d\lambda \, dJ$.

The first term in the parentheses in Eq. (10) corresponds to the flat space. Therefore, according to Eqs. (7) and (8),

$$S(R) - S(0) = \frac{\pi}{4} \frac{V}{2\pi} \frac{1}{4\pi^2 a^2} \int_0^\infty dk \int_{-\infty}^{+\infty} d\omega \, f(\lambda)[c_0 \, \mathrm{sign}(\lambda - M_0^2)]$$

$$+ c_1 \, \mathrm{sign}(\lambda - M_1^2) \quad (11)$$

Integral (11) diverges if the particles M_0 and M_1 are real; thus we formally assume, by taking $M \sim k_0$, that the "ghost" particles exist, with weight in (11) negative, i.e.,

$$c_0 M_0^2 + c_1 M_1^2 = 0 \quad (12\mathrm{a})$$

$f(\lambda)$ is the cutting function, which we assume to be "even," i.e.,

$$f(\lambda) = \begin{cases} 1 & \text{at } |\lambda| < k_0^2 \\ 0 & \text{at } |\lambda| > k_0^2 \end{cases} \quad (12\mathrm{b})$$

If $f(\lambda)$ is not "even," i.e.,

$$f(\lambda) = 1 \quad \text{if } -k_1^2 < \lambda < k_0^2$$

$$\frac{k_0^2 - k_1^2}{2} = M_0^2 \quad (12\mathrm{c})$$

FIELD THEORY AND ELEMENTARY PARTICLES

(11) converges without the introduction of "ghost" particles. Performing the integration of the improper integral (11) in the variables k, ω, we find

$$S(R) - S(0) = \frac{V c_0 M_0^2 \ln |c_0/c_1|}{32\pi^2 a^2} \tag{13}$$

Comparing Eqs. (13) and (1), we find

$$G = \frac{12\pi}{c_0 M_0^2 \ln |c_0/c_1|} \tag{14}$$

Thus $M_0 \sim 10^{28}$ eV, $|M_1| > M_0$.

The possibility of neglecting the Lagrangian of the free electromagnetic field in charge-renormalization calculations was first pointed out in Refs. 3 and 4.

Our approach to gravitation is rather similar to the approach to quantum electrodynamics used in Ref. 4. Formula (14) of the present work is similar to their formula for e_0^2, determining the effective number of charged fields [see Eq. (A4)].

The author is grateful to Ya. B. Zel'dovich for the discussion which initiated the present investigation, as well as for allowing him to read Refs. 2 and 5 prior to publication and for useful advice.

In Ref. 5, Zel'dovich develops the idea of exclusion of the initial Lagrangian of the free field from electrodynamics. In the Appendix we investigate the same problem by the method of functional integration, along the same lines as in our treatment of gravitation.

APPENDIX

Let us calculate the variation of action of spin zero charged bosons, as well as of spin $\frac{1}{2}$ charged fermions by switching on the electromagnetic field.

In the presence of the field the eigenfunctions of the scalar field can be found from the equation

$$\left[\left(\frac{\partial}{\partial x_0} - ie_0 A_0\right)^2 - \left(\frac{\partial}{\partial \xi_\alpha} - ie_0 A_\alpha\right)^2 + \lambda_i\right] \phi_i = 0 \tag{A1}$$

e_0 being the elementary charge.

In the case of the fermion, ϕ_i is the two-component spinor, and moreover, the term $\sigma(H + iE)$ is to be added.

We assume that the variation of the spectral density $P = dN/d\lambda$ for the quantities λ_i due to switching the field on can be represented by the equation

$$P(A) - P(0) = \frac{Be_0^2}{\lambda} \int dx\, (E^2 - H^2) \tag{A2}$$

Eq. (A2) expresses a purely mathematical statement. We have not proved it in general; however, its validity seems to be very plausible because it holds for the case of a constant uniform magnetic field. For this case we find

$$B_0 = \frac{1}{48\pi^3} \quad \text{(scalar field)}$$

$$B_i = -\frac{1}{12\pi^3} \quad \text{(spin } \tfrac{1}{2}\text{)}$$

Using these values of the constants as well as Eqs. (A2), (7), and (8) for the cutting function (12b), we can easily calculate the principal value of the integral above:

$$S(A) - S(0) = \pi e_0^2 \sum_j |B| \ln \frac{k_0}{m_j} \int dx\, (E^2 - H^2) \tag{A3}$$

The sum is taken over various kinds of charged particles in a similar way (Ref. 4).

According to the idea due to Ya. B. Zel'dovich [5], this expression is equal to the action of electromagnetic field $(1/8\pi)\int dx\, (E^2 - H^2)$. If all the particles in Eq. (A3) are $\tfrac{1}{2}$ spin fermions, we find:

$$e_0^2 = \frac{3\pi}{2 \sum_j \ln(k_0/m_j)} \tag{A4}$$

This formula coincides with the one obtained by Landau and Pomeranchuk [4] by calculating the vacuum polarization. For scalar particles this result corresponds to that of Gorkov and Khalatnikov [6].

REFERENCES

1. E. M. Lifshits, *Sov. Phys. JETP* 2:94 (1954). In this work the molecular attraction of condensed bodies was calculated as due to the change of the electromagnetic fluctuation spectrum. As was pointed out in this paper, the special case of attraction of metals had been previously investigated by H. B. G. Casimir, *Proc. Nederl. Akad. Wetensch.* 51:793 (1948).
2. Ya. B. Zel'dovich, *ZhETF Pis'ma* 6:922 (1967); *JETP Lett.* 6:345 (1967), trans.
3. E. S. Fradkin, *Dokl.* 98:47 (1954); *100*:897 (1955).

4. L. D. Landau, I. Ya. Pomeranchuk, *Dokl. 102*:489 (1955); trans. in Landau's Collected Papers (D. ter Haar, ed.), Pergamon Press, 1965.
5. Ya. B. Zel'dovich, *ZhETF Pis'ma 6*:1233 (1967).
6. L. P. Gorkov, I. M. Khalatnikov, *Dokl. 104*:197 (1955).

Paper 16

Spectral Density of Eigenvalues of the Wave Equation and Vacuum Polarization

The effective Lagrange function of vacuum polarization is expressed in terms of the spectral density of the eigenvalues of the wave equation and the related five-dimensional Green's function introduced by Fock in his fifth-coordinate method. The method can be used for arbitrarily strong external fields, but the interaction of the vacuum fields is neglected. The vacuum polarization by the gravitational and electromagnetic fields is calculated.

I. INTRODUCTION

This paper augments methodologically and mathematically my papers (Refs. 1 and 2) in which I put forward a conjecture about the Lagrangian of the gravitational and the electromagnetic field (in the second case the basic idea is due to Pomeranchuk and Landau, Fradkin, and Zel'dovich [3]). In its simplest form, the conjecture states that the Lagrange function of boson fields (gravitational, electromagnetic, and meson) is generated by vacuum polarization effects of fermions. In this work the term vacuum polarization is used in a wider sense than usual—the Lagrange functions of free boson fields and even the cosmological constant are attributed to polarization.

In this paper, a method of calculating the effective polarization Lagrange function is developed on the basis of the concept of the spectral density of the wave equation (Sec. II). In Sec. III, the spectral density is related to the Green's function defined in five-dimensional space (physical space augmented by a fifth auxiliary coordinate). The auxiliary fifth

Source: Спектральная плотность собственных значений волнового уравнения и поляризация вакуума, Теор. и мат. физика 23:178-190 (1975).   Reprinted from *Theor. Math. Phys.* 23:435–444 (1976), with permission from Plenum Publishing Corporation.

coordinate ("proper time") was first introduced by Fock in 1937 [9]. This method was further developed by Schwinger and others [10]. Here, the general expression (25) for the effective polarization Lagrange function is derived differently.

In Sec. IV, the general method is applied to the gravitational field. In a model theory a formal cutoff of divergent integrals is used to find an expression for the gravitational constant, which has the correct sign ($G > 0$). In Sec. V, the method is illustrated for the example of the electromagnetic field, and the well-known expressions for the vacuum polarization by the electromagnetic field are again obtained. The signature of the metric tensor is $(+---)$, and gravitational units, in which $G = \hbar = c = 1$, are used.

II. SPECTRAL DENSITY OF THE EIGENVALUES OF THE WAVE EQUATION

Ref. 2 contains a sketch of the idea behind the new method of calculating vacuum polarization. So as to make the exposition comprehensive, some of this section [Eqs. (1) to (5)] repeats these ideas with some necessary refinements.

We consider the effect of vacuum polarization by an external field $\psi(x)$, which we assume fixed. The elementary fields with which ψ interacts will be denoted by $\phi_1, \phi_2, \ldots, \phi_j, \ldots$. We ignore the interaction of the fields ϕ with one another. This is the main assumption of the paper, and is equivalent to the restriction to single-loop diagrams. Because of this assumption, the paper does not have great physical significance but is rather methodological, or mathematical, in character. To illustrate the idea of Ref. 2 in the simplest way possible, the ϕ's will be assumed to be neutral scalar fields, and the field ψ a given neutral tensor g_{ik} or vector A_i field; g_{ik} is the metric tensor of the gravitational field and A_i the potential of the electromagnetic field.

Suppose that the field $\psi(x)$ is defined in some four-dimensional volume V. The total action of the fields ϕ_j in this volume is a functional $S(\psi)$. We denote the value of the functional for $\psi = 0$ (vacuum value) by $S(0)$. Obviously, $S(\psi) - S(0)$ is the vacuum polarization effect due to the field ψ. By the conjecture, this difference is the effective action of the field ψ. The functional $S(\psi) = \Sigma S_j$ is the sum of the functionals for the individual fields ϕ_j. We calculate one of these terms (omitting for brevity the subscript j). We expand ϕ in a series in eigenfunctions of the wave equation (in the simplest case, this is simply an expansion in a four-dimensional Fourier series):

$$\phi = \Sigma \, z_i \phi_i \tag{1}$$

$$\Box_\psi \phi_i + (m^2 + \Lambda_i)\phi_i = 0$$

$$\int dx \, (-g)^{1/2} \phi_i \phi_{i'}^+ = \delta_{ii'} \tag{2}$$

Here the symbol $\Box_\psi$ stands for the generalized d'Alembertian in the presence of the given field ψ. If $\psi = 0$, than $\Box_0 = \partial^2/\partial t^2 - \partial^2/\partial x_1^2 - \partial^2/\partial x_2^2 - \partial^2/\partial x_3^2$. If $\psi = A_i$ is the electromagnetic field, then $\nabla \to \nabla_\psi = \nabla - ieA$. If $\psi = g_{ik}$, then $\Box_\psi$ is the Beltrami operator[†]; ψ_i^+ is an eigenfunction of the adjoint equation; Λ_i is an eigenvalue of the wave equation, and every Λ_i is a functional of ψ. If $\psi = 0$, $\Lambda_i = \omega_i^2 - k_i^2 - m^2$. We assume that the classical action for the field ϕ is

$$S_{cl} = \sum_i S_i \tag{3}$$

$$S_i = \frac{z_i^2}{2} \Lambda_i$$

This accords with the classical equations of motion. We find the action of the quantum fluctuations as the phase of the functional integral in the case of variation of ϕ:

$$S = \arg \int (\delta\phi) e^{iS_{cl}} = \Sigma \arg \int_{-\infty}^{+\infty} dz_i \, e^{iz_i^2 \Lambda_i/2} = \frac{\pi}{4} \Sigma \, \text{sign} \, \Lambda_i \tag{4}$$

The integral with respect to dz_i is calculated by the change of variables

$$iz_i^2 = -\zeta_i^2 \, \text{sign} \, \Lambda_i$$

$$dz_i = d\zeta_i \, e^{i(\pi/2)\text{sign}\Lambda_i}$$

Generalizing (4) to the presence of spinor fields ϕ_j and taking into account the statistical weight g_j, we obtain the general expression

$$S(\psi) = \frac{\pi}{4} \sum_j g_j C_j \sum_i \text{sign} \, r\Lambda_{ij} \tag{5}$$

Here

$$C_j = \begin{cases} +1 & \text{if } \phi_j \text{ is a boson field} \\ -1 & \text{if } \phi_j \text{ is a fermion field} \end{cases} \tag{5a}$$

[†] Generally speaking, the equation for the scalar ϕ may contain an additional term $(-R\phi/6$, where R is the trace of the Ricci tensor), which in the case $m = 0$ makes the theory conformally invariant. For the discussion of the basic ideas, these details are not important.

The factor C_j takes into account the fact that for spinor fields the contribution to the action has the opposite sign. In the sum (5) we have formally introduced a convergence factor of unknown physical origin. The "cutoff" weight function is

$$\text{sign } r\Lambda = \begin{cases} \text{sign } \Lambda & \text{for } |\Lambda| < \Lambda_0 \\ 0 & \text{for } |\Lambda| > \Lambda_0 \end{cases} \tag{5b}$$

or

$$\text{sign } r\Lambda = e^{-|\Lambda|/\Lambda_0} \text{ sign } \Lambda \tag{5c}$$

where Λ_0 is the square of the cutoff mass. We assume $\Lambda_0 \sim 1$ in gravitational units.

For the following discussion we consider not the particular form of the function sign $r\Lambda$, but a sum with arbitrary function $\Phi(\Lambda)$:

$$\Sigma_\Phi = \sum_i \Phi(\Lambda_i) \tag{6}$$

The sum Σ_Φ diverges, since in any interval $(\Lambda, \Lambda + d\Lambda)$ there are infinitely many eigenvalues Λ_i. For example, in the case $\psi = 0$, taking V to be a parallelepiped, we find that the points k_{i0}, k_{i1}, k_{i2}, k_{i3} form an infinite periodic four-dimensional lattice. Between the two hyperboloids $\Lambda = \text{const}$ and $\Lambda + d\Lambda = \text{const}$ there is an infinite volume containing infinitely many sites of the lattice. The situation is similar in the general case.

We now introduce the concept of "conditional covergence" of the sum Φ. For simplicity, we restrict ourselves to the case when the volume V is topologically equivalent to a four-dimensional cube. Deforming V continuously into a cube L^4 and taking the limit as $\psi \to 0$, we map the functions ϕ_i into functions of the form $\exp(2\pi i/L)(n_0 x_0 - n_1 x_1 - n_2 x_2 - n_3 x_3)$. We define an invariant of the deformation process: $J(i) = n_0^2 + n_1^2 + n_2^2 + n_3^2$. The sum Σ_Φ converges conditionally if the following limit exists:

$$\lim_{J_0 \to \infty} \sum_i e^{-J(i)/J_0} \Phi(\Lambda_i) \overset{\text{def}}{=} \Sigma_\Phi \tag{7}$$

Other equivalent definitions are possible.

The sum (6) is conditionally convergent if the summability conditions

$$\int_{-\infty}^{+\infty} d\Lambda \, \Phi(\Lambda) = 0 \tag{8}$$

$$\int_{-\infty}^{+\infty} d\Lambda \, \Lambda \Phi(\Lambda) = 0 \tag{9}$$

FIELD THEORY AND ELEMENTARY PARTICLES 183

are satisfied and the function $\Phi(\Lambda)$ decreases sufficiently rapidly as $\Lambda \to \infty$; the idea behind the proof of this assertion is sketched in the next section.

We now define the spectral density $P(\Lambda)$ of the eigenvalues of the wave equation by requiring

$$\lim_{\varepsilon \to 0} \frac{1}{\varepsilon^k} \left[\sum_i \Phi_k(\varepsilon \Lambda_i) - \int_{-\infty}^{+\infty} \Phi_k(\varepsilon \Lambda) P(\Lambda) \, d\Lambda \right] = 0 \qquad (10)$$

The functions Φ_k satisfy certain conditions, which depend on k. For $k = 1, 2$, we assume that $\Phi(\varepsilon \Lambda)/\varepsilon$ does not tend to infinity as $\delta \to 0$.

By virtue of the conditions (8) and (9), the function $P(\Lambda)$ is not defined uniquely by (10), but only up to the addition of an arbitrary linear function of Λ. The integral $\int_{-\infty}^{+\infty} \Phi(\Lambda) P(\Lambda) \, d\Lambda$ does not change under the transformation

$$P(\Lambda) \to P(\Lambda) + A\Lambda + B \qquad (11)$$

Representing the function $P(\Lambda)$ as the series

$$P(\Lambda) = C_0 \lambda \ln \frac{|\lambda|}{\lambda_0} + C_1 \ln \frac{|\lambda|}{\lambda_1} + \frac{C_2}{\lambda} + \frac{C_3}{\lambda^2} + \cdots \qquad (12)$$

where $\lambda = m^2 + \Lambda$, we determine the coefficients C_0, C_1, etc., successively from (10). These coefficients do not depend on the form of the function Φ. The coefficients λ_0 and λ_1 are arbitrary in accordance with (11).

III. GREEN'S FUNCTION METHOD IN FIVE-DIMENSIONAL SPACE

In Ref. 4, McKean and Singer considered the Helmholtz equation in n-dimensional Riemannian space with definite metric (we change their notation slightly):

$$\Delta \phi_i + \lambda_i \phi_i = 0 \qquad (13)$$

They show that the sum

$$\Sigma(\tau) = \sum_i e^{-i\lambda_i \tau} \qquad (14)$$

can be calculated by means of a Green's function in an auxiliary space of $n + 1$ dimensions. Here, the Green's function $G(x_0, x_1, \tau)$ is a normalized singular solution of the equation

$$\Delta G = \frac{\partial G}{\partial \tau} \qquad (15)$$

In Ref. 4 it is shown that

$$\Sigma(\tau) = \int dx \, (g)^{1/2} G(x, x, \tau) \qquad x = x_0 = x_1 \qquad (16)$$

McKean and Singer point out that M. Kac was one of the originators of the basic idea.

We apply here a similar method to find the density function $P(\Lambda)$ in the case of real physical space, i.e., for the wave equation (1). Denoting the auxiliary fifth variable by the letter l, we write down by analogy with (15) an equation for the five-dimensional Green's function:

$$\Box_\psi G = i \frac{\partial G}{\partial l} \qquad (17)$$

An equation of the type (17) was first introduced by Fock [9]. The Green's function G depends on the nine variables t_0, x_0, y_0, z_0 (abbreviated x_0), t_1, x_1, y_1, z_1 (abbreviated x_1), and $l = l_1 - l_0$. If ϕ has many components, G also depends on the discrete numbers of the initial and final state with respect to l, i.e., it is a matrix $G_{m_0 m_1}$ (the use of the same notation for the discrete variable and the field mass m should not cause confusion). The Green's function defined by (17) satisfies an integral relation similar to (16):

$$\Sigma(l) = \sum_i e^{i\lambda_i l} = \int dx \, (-g)^{1/2} \mathrm{Tr} \, G_{m_0 m_1}(x, x, l) \qquad x = x_1 = x_0 \qquad (18)$$

where λ_i are the eigenvalues of the wave equation

$$\Box_\psi \phi_i + \lambda_i \phi_i = 0 \qquad \lambda_i = \Lambda_i + m^2$$

Equation (18) is proved by representing the Green's function as the conditionally convergent sum [see (7) above]

$$G_{m_0 m_1}(x_0, x_1, l) = \lim_{J_0 \to \infty} \sum_i \exp\left(-\frac{J}{J_0} + il\lambda_i\right) \phi_i(x_1, m_1) \phi_i^+(x_0, m_0)$$
(19)

The function defined by (19) satisfies Eq. (17) and goes over into the four-dimensional δ-function as $l \to 0$ on account of the orthogonality relations

$$\lim_{J_0 \to \infty} \sum_i e^{-J/J_0} \phi_i(x_1, m_1) \phi_i^+(x_0, m_0) = \delta(x_0 - x_1) \delta_{m_0 m_1}$$

Setting $x_1 = x_0 = x$ and $m_1 = m_0$ in (19), we integrate over x and sum over m. We arrive at (18), in which the sum over i is also understood in the sense of (7). The fact of convergence of the sum accords with the fact that the function $e^{i\lambda l}$ satisfies the conditions (8) and (9) when $l \neq 0$. We

FIELD THEORY AND ELEMENTARY PARTICLES

now show how these conditions are related to the convergence of the sum (7) for arbitrary weight function $\Phi(\Lambda)$. We set

$$\Phi(\lambda) = \int_{-\infty}^{+\infty} f(l) e^{i\lambda l} \, dl$$

$$f(l) = \frac{1}{2\pi} \int_{-\infty}^{+\infty} \Phi(\lambda) e^{-i\lambda l} \, d\lambda$$

It follows from (8) and (9) that $f(0) = (df/dl)(0) = 0$, which allows convergence of the integral $I = \int dl \, dx \, \text{Tr} \, G(x, x, l) f(l)$ in the neighborhood of $l = 0$ despite $G(l)$'s having a singularity at $l = 0$ of the form $\alpha/l|l| + \beta/|l|$ [see (23) below]. The integral converges in this case if there are additional but hardly restrictive conditions on the function $\Phi(\Lambda)$ [for example, if $\Phi(\Lambda)$ is bounded in modulus and decreases as $\lambda \to \pm\infty$ in such a way that its integral is absolutely convergent].

It follows from (18) that

$$I = \int dl \lim_{J_0 \to \infty} \sum_i \exp\left(-\frac{J}{J_0} + i\lambda_i l\right) f(l)$$

or, reversing the order of summation and integration,

$$\lim_{J_0 \to \infty} \sum_i e^{-J/J_0} \Phi(\lambda_i) = I$$

Thus, the sum (7) is reduced to the finite value I.

We now introduce the local spectral density $\rho_m(\lambda, x)$ of the wave equation, which is related to the integral density $P(\lambda)$ by

$$\sum_m \int dx \, (-g)^{1/2} \rho_m(\lambda, x) = P(\lambda) \tag{20}$$

(recall that $\lambda = m^2 + \Lambda$).

From (18) and (20),

$$\int_{-\infty}^{+\infty} \rho(\lambda, x, m) e^{i\lambda l} d\lambda = G_{mm}(x, x, l) \tag{21}$$

We find the form of G in flat space for $\psi = 0$. We have

$$G_0 = G_t G_x G_y G_z$$

$$G_t = a_t \frac{1}{(4\pi l)^{1/2}} e^{-it^2/4l}$$

$$G_x = \frac{a_x}{(4\pi l)^{1/2}} e^{ix^2/4l}$$

and similarly for y and z. The normalization coefficients can be determined from the conditions $\int_{-\infty}^{+\infty} dt\, G_t(t, t, l) = 1$, etc., separately for $l > 0$ and $l < 0$.

We find

$$G_0 = -\frac{i \operatorname{sign} l}{(4\pi l)^2} e^{-i(t^2-x^2)/4l} \tag{22}$$

Writing $\rho(\lambda)$ and $G(l)$ as series, we have

$$\rho_m = a_0 \lambda \ln\frac{|\lambda|}{\lambda_0} + a_1 \ln\frac{|\lambda|}{\lambda_1} + a_2 \lambda^{-1} + a_3 \lambda^{-2} + \cdots \tag{23}$$

$$G_{mm} = \operatorname{sign} l(A_0 l^{-2} + A_1 l^{-1} + A_2 + A_3 l + \cdots)$$

$$a_0 = \frac{iA_0}{\pi} = \frac{1}{16\pi^3} \qquad a_1 = -\frac{A_1}{\pi} \qquad a_2 = -\frac{iA_2}{\pi}$$

$$a_3 = -\frac{A_3}{\pi} \qquad a_4 = \frac{2iA_4}{\pi}, \ldots$$

This connection between a and A follows from (21). It follows from (5) that the increment of the effective Lagrangian due to the vacuum polarization of the particles of the field ϕ in the presence of some external field ψ is

$$\Delta \mathcal{L}_j = \frac{\pi C_j}{4} \sum_m \int_{-\infty}^{+\infty} d\lambda \operatorname{sign} r\Lambda\, (\rho_\psi - \rho_0) \tag{24}$$

Applying to (24) and (20) the convolution theorem from the theory of Fourier integrals, we represent $\Delta \mathcal{L}_j$ as an integral with respect to the auxiliary variable l,

$$\Delta \mathcal{L}_j = \frac{\pi C_j}{4} \int_{-\infty}^{+\infty} dl\, Z(l) R_\psi(l)$$

$$R_\psi(l) = \frac{1}{\sqrt{2\pi}} \int_{-\infty}^{+\infty} \sum_m (\rho_\psi - \rho_0) e^{i\lambda l}\, d\lambda = \frac{1}{\sqrt{2\pi}} (\operatorname{Tr} G_\psi - \operatorname{Tr} G_0)$$

$$Z(l) = \frac{1}{\sqrt{2\pi}} \int_{-\infty}^{+\infty} \operatorname{sign} r(\lambda - m_j^2) e^{-i\lambda l}\, d\lambda = -\frac{2i}{\sqrt{2\pi}} e^{-im_j^2 l} \operatorname{Reg}\{1/l\}$$

where $\operatorname{Reg}\{1/l\}$ is the function "cut off" for $|l| \lesssim 1/\Lambda_0$ [using (5c), we have $\operatorname{Reg}\{1/l\} = l/l^2 + \Lambda_0^{-2}$]. Substituting this, we find

FIELD THEORY AND ELEMENTARY PARTICLES

$$\Delta\mathcal{L}_j = -\frac{iC_j}{4}\int_{-\infty}^{+\infty} dl \,\text{Reg}\,\{1/l\}e^{-im_j^2 l}\,(\text{Tr}\,G_\psi - \text{Tr}\,G_0)$$

$$= \frac{C_j}{2}\int_{1/\Lambda_0}^{\infty}\frac{dl}{l}\,\text{Im}\,[e^{-im_j^2 l}(\text{Tr}\,G_\psi - \text{Tr}\,G_0)]$$

$$= \frac{C_j}{2}\int_{1/\Lambda_0}^{\infty}\frac{dl}{l}\,[\cos m_j^2 l(\text{Im}\,\text{Tr}\,G_\psi - \text{Im}\,\text{Tr}\,G_0)$$

$$- \sin m_j^2 l\,\text{Re}\,\text{Tr}\,G_\psi] \tag{25}$$

IV. COSMOLOGICAL CONSTANT AND POLARIZATION OF THE VACUUM BY THE GRAVITATIONAL FIELD

The Lagrange function $\mathcal{L}_0$ of flat space for $\psi = 0$ is (up to sign) Einstein's cosmological constant (compare this with the formulation of the problem and the calculations of Zel'dovich [5]). If the summation conditions (8) and (9) are satisfied, then on the basis of (5),

$$\mathcal{L}_0 = \frac{\pi}{4}\sum_j g_j C_j \sum_i \text{sign}\,r\Lambda_{ij}$$

$$= \frac{\pi}{4}\int_{-\infty}^{+\infty} d\lambda \sum_j C_j \rho_{0j}(\lambda)\,\text{sign}\,r(\lambda - m_j^2) \tag{26}$$

where $\rho_0 = g_j[\lambda \ln (|\lambda|/\Lambda_0)]/16\pi^3$.

Recall that $C_j = \pm 1$ for bosons (fermions). The summability condition (8) is satisfied automatically [since $\int \text{sign}\,r(\Lambda)\,d\lambda = 0$], and the condition (9) leads to

$$\Sigma\,g_j C_j = 0 \tag{27}$$

The integral (26) subject to (27) can be calculated by an equation of the type (25):

$$\mathcal{L}_0 = -\frac{1}{2}\int_{1/\Lambda_0}^{\infty}\frac{dl}{l}\sum_j g_j C_j \frac{\cos m_j^2 l}{16\pi^2 l^2} \tag{28}$$

To logarithmic accuracy,

$$\mathcal{L}_0 = \frac{1}{64\pi^2}\Sigma\,g_j C_j m_j^4 \ln\frac{\Lambda_0}{m_j^2} \tag{29}$$

In reality, it is well known that $\mathcal{L}_0$ (up to sign it is Einstein's cosmological constant) is either zero or extremely small. In a model theory of noninteracting particles $\mathscr{L}_0$ can vanish as a result of the composition of the contributions of bosons and fermions. In a more realistic theory with allowance for interaction and spontaneous symmetry breaking the vanishing of $\mathscr{L}_0$ must be regarded as a physical condition imposed on the constants in the unrenormalized Lagrangian.[†]

In curved Riemannian space (i.e., in a space in which the Riemann tensor R_{iklm} is nonzero), $\rho(\lambda)$, $G(l)$, and $\mathscr{L}$ are changed (polarization of the vacuum by the gravitational field).

We represent the Green's function in the form

$$G_{Rj} = G_{0j}[1 - iQRl - (U_j R^2 + V_j R^{ik} R_{ik} + W_j R^{iklm} R_{iklm} + Y\Box R)l^2 + \cdots] \tag{30}$$

The coefficients Q, U, etc., are found in the Appendix for scalar fields ϕ_j.

Substituting into (25) and using $G_{0j} = -ig_j \operatorname{sign} l/16\pi^2 l^2$, we obtain the first-order correction

$$\Delta\mathscr{L}_{1j} = -\frac{RQ_j C_j g_j}{2} \int_{1/\Lambda_0}^{\infty} \frac{dl}{l} \frac{\sin m_j^2 l}{16\pi^2 l} \tag{31}$$

The similar expression for the second-order correction is

$$\Delta\mathscr{L}_{2j} = \frac{U_j R^2 + V_j R^{ik} + \cdots}{2} \int_{1/\Lambda_0}^{\infty} \frac{dl}{l} \frac{\cos m_j^2 l}{16\pi^2} \tag{32}$$

Calculation of the integrals in (31) and (32) to logarithmic accuracy gives

$$\Delta\mathscr{L}_{1j} = \frac{RQ_j C_j g_j}{32\pi^2} m_j^2 \ln \frac{\Lambda_0}{m_j^2} \tag{31a}$$

$$\Delta\mathscr{L}_{2j} = \frac{U_j R^2 + \cdots}{32\pi^2} \ln \frac{\Lambda_0}{m_j^2} \tag{32a}$$

Equating $\sum_j \Delta\mathscr{L}_{1j} = -R/16\pi G$ in accordance with the conjecture, we find that the gravitational constant is

$$G = \frac{-2\pi}{\sum_j Q_j C_j g_j m_j^2 \ln(\Lambda_0/m_j^2)} \tag{33}$$

Thus, in the model theory with formal cutoff at $|\Lambda| \sim \Lambda_0$ we have found the correct sign $G > 0$ with allowance for $Q_j C_j \geq 0$. Equation (33) gives the correct numerical value of G (equal to 1 in the chosen units) if the

---

[†] $\mathscr{L}_0 = 0$ in theories with supersymmetry and possibly in theories with spontaneous breaking of supersymmetry (communication of V. I. Ogievetskii).

mass spectrum of the elementary fields extends to $m_j \sim G^{-1/2}$. The expression (32a) diverges for particles with rest mass $m_j = 0$ (neutrino, photon, graviton; the last case requires special treatment).

The quadratic correction $\mathscr{L}_{2j}$ for gravitons has been considered by De Witt [6]. He assumed that for particles of zero mass the logarithmic divergence is cut off in the infrared limit at lengths that depend on the characteristic scale L of the problem. Since our method does not require an expansion in a series in powers of the curvature tensor, it automatically leads to a cutoff of the infrared divergence at $l \sim L^2$.

Let us demonstrate this for the example of a scalar field without the conformal term in the equation of motion. We consider a space with the metric

$$ds^2 = dl^2 - dr^2 - L^2 \sinh^2 \frac{r}{L} (d\vartheta^2 + \sin^2 \vartheta \, d\phi^2)$$

Applying the method described in the Appendix, we find $G(0, 0, l) = G_0(0, 0, l)e^{-il/L^2}$. Substituting into (25), we have

$$\Delta \mathscr{L} = \frac{1}{32\pi^2} \int_{1/\Lambda_0}^{\infty} dl \, \frac{1 - \cos(l/L^2)}{l^3} = \frac{\ln(L^2\Lambda_0)}{64\pi^2 L^4} \tag{34}$$

The coefficient in the expression with allowance for $U = 1/72$ and $6/L^2 = R$ corresponds to Eq. (32a) with the cutoff $l_{max} = L^2$, $l_{min} = 1/\Lambda_0$.

V. VACUUM POLARIZATION BY THE ELECTROMAGNETIC FIELD

Another example of application of the general method is the vacuum polarization by a given electromagnetic field. We assume that the vectors **E** and **H** do not depend on the coordinates. If either of the invariants $J_1 = \mathbf{E}^2 - \mathbf{H}^2$ or $J_2 = (\mathbf{EH})^2$ is nonzero, there exists a Lorentz transformation as a result of which $\mathbf{E} \to \mathbf{E}_0$, $\mathbf{H} \to \mathbf{H}_0$, so that $\mathbf{E}_0 \| \mathbf{H}_0$ (and to be specific we take them along the x axis). The vector potential in this frame of reference has components $A_t = -xE_0/2$, $A_x = tE_0/2$, $A_y = -zH_0/2$, $A_z = yH_0/2$. We calculate $\Delta \mathscr{L}$ at the point (0, 0, 0, 0). We consider first a complex scalar field ϕ. The equation for the Green's function,

$$\Box_A G_A = i \frac{\partial G_A}{\partial l} \quad \Box_A = \left(\frac{\partial}{\partial t} - ieA_t\right)^2 - \left(\frac{\partial}{\partial x} - ie\mathbf{A}\right)^2 \tag{35}$$

has the solution $G_A = G_1(0, 0, x, t, l)G_2(0, 0, x, z, l)$. Substituting into (35),

$$G_1 = \frac{1}{4\pi l} \exp\left[-i(t^2 - x^2)\alpha(l) + \beta(l)\right]$$

$$G_2 = -\frac{i}{4\pi l} \exp\left[i(y^2 + z^2)\gamma(l) + \beta(l)\right] \operatorname{sign} l$$

we find

$$\alpha = \frac{eE_0}{4} \coth(eE_0 l)$$

$$\beta = \ln \frac{eE_0 l}{\sinh(eE_0 l)}$$

$$\gamma = \frac{eH_0}{4} \cot(eH_0 l)$$

$$\delta = \ln \frac{eH_0 l}{\sin(eH_0 l)}$$

Hence, the Green's function of the scalar complex field for $x_1 = x_0 = (0, 0, 0, 0)$ is

$$G_A = -\frac{i \operatorname{sign} l}{(4\pi)^2} \frac{eE_0}{\sinh(eE_0 l)} \frac{eH_0}{\sin(eH_0 l)} \tag{36}$$

For a charged spinor field, Eq. (35) must be regarded as a four-row equation with the substitution (Σ_x and α_x are Dirac matrices)

$$\Box_A \to \Box'_A = \delta_{m_0 m_1} \Box_A + \Sigma_x eH_0 + i\alpha_x eE_0 \tag{36a}$$

$$G'_A(0, 0, l) = \delta_{m_0 m_1} G_A \exp l(-ieH_0\Sigma + eE_0\alpha)$$

The contribution of the spinor with mass $m = m_j$ on the basis of (36a) and (25) is

$$\Delta \mathscr{L}_{jA} = \frac{1}{8\pi^2} \int_{1/\Lambda_0}^{\infty} \frac{dl}{l} \cos m^2 l \left[\frac{e^2 E_0 H_0}{\tan(eH_0 l)\tanh(eE_0 l)} - \frac{1}{l^2}\right] \tag{37}$$

(cf. Ref. 8).

Expanding coth and cot in series[†] and expressing the polynomials in E_0 and H_0 in terms of the invariants J_1 and J_2, we obtain for the expression in the square brackets in (37)

[†]This is done conveniently on the basis of the formulas $\coth(2x) = \frac{1}{2}(\coth x + 1/\coth x)$, $\cot(2x) = \frac{1}{2}(\cot x - 1/\cot x)$ by the method of undetermined coefficients.

FIELD THEORY AND ELEMENTARY PARTICLES 191

$$[\quad] = \frac{e^2 J_1}{3} - e^4 \left(\frac{J_1^2}{45} + \frac{7J_2}{45}\right) l^2 + e^6 \left(\frac{2}{945} J_1^3 + \frac{13}{945} J_2 J_1\right) l^4 + \cdots$$

After integration of (37),

$$\Delta \mathscr{L}_{jA} = \frac{e^2 J_1}{24\pi^2} \ln \frac{\Delta}{m^2} + \frac{e^4(J_1^2 + 7J_2)}{360 m^4 \pi^2} + \frac{e^6(2J_1^3 + 13 J_2 J_1)}{1260 \pi^2 m^8} + \cdots \quad (38)$$

The sum of the logarithmic terms for all charged particles must in accordance with the conjecture be $\Delta \mathscr{L}_1 = (\mathbf{E}^2 - \mathbf{H}^2)/8\pi$, and from this "sum rule" one can find the fine structure constant e^2 (Landau and Pomeranchuk [3]). The second term in (38) is the vacuum polarization found by Weisskopf in 1936 [7]. The third term describes six-photon processes.

VI. CONCLUSIONS

We have rederived the expressions of the proper-time method of Fock and Schwinger based on the concept of the spectral density of a wave equation. We have found an expression for the polarization by the gravitational field of the vacuum of scalar, spinor, and vector particles to terms quadratic in the components of the curvature tensor; for zero-mass particles we have obtained an expression that does not use an expansion in powers of the curvature tensor and does not contain an infrared divergence. The method has also been illustrated by the example of the electromagnetic field. The method can be readily generalized to any processes that can be described by single-loop diagrams. For example, the method is fully applicable to the calculation of the effective density of the Lagrange function of boson fields and also fermion fields with nonzero masses and charges, and to the calculation of the radiative corrections to the magnetic moment of particles with spin in an arbitrarily strong external field (i.e., to the calculation of not only the "intrinsic" magnetic moment, but also the polarizability), to the calculation of the effective Lagrangian of vector fields of the Yang–Mills type, etc. However, all these calculations presuppose a restriction to single-loop diagrams if the method is used unmodified. How to extend the method to diagrams of more general form is not clear.

I thank the participants of the Theoretical Seminar of the P. N. Lebedev Physics Institute for discussing a first version of the work in June 1970, and also Ya. B. Zel'dovich for numerous discussions of the basic ideas. His paper on the cosmological constant [5] and the Lagrangian of the electromagnetic field [3] were important stimuli for this work. I thank I. M. Gel'fand for discussion and for drawing my attention to the work of McKean and Singer [4], and also giving me a photocopy of Ref. 4.

APPENDIX

Following McKean and Singer's method, we find the Green's function of n-dimensional Riemannian space with definite metric for a scalar, a spinor, and a vector field.

We represent the Green's function of Eq. (15) in the form

$$G_R = G_0[1 + QR\tau + (UR^2 + VR^{ik}R_{ik} + WR^{iklm}R_{iklm} + Y\Box R)\tau^2 + \cdots] \quad (A.1)$$

The coefficients Q, U, etc., in this equation do not, according to Ref. 4, depend on the dimensionality n of the space. Therefore, all the coefficients except Y can be found by studying the solutions for three spaces of different dimensionality and constant curvature (for example, a sphere S^2, a hypersphere S^3, and supersphere S^4).

For each of these spaces one finds the eigenfunctions ϕ_i, determines g_i and λ_i, and forms the sum

$$G(\tau) = G_0(\tau)(1 + a_1\tau + a_2\tau^2 + \cdots) = \frac{1}{V}\sum g_i e^{-\lambda_i \tau} \quad (A.2)$$

The sums are calculated by means of the asymptotic series

$$\sum_0^\infty f(n) = \int_0^\infty f(n)\,dn + \frac{f(0)}{2} - \frac{f'(0)}{12} + \frac{f'''(0)}{720} + \cdots \quad (A.3)$$

We denote the coefficients for a sphere by $a_1^{(2)}$ and $a_2^{(2)}$ and similarly for the hypersphere and supersphere.

We find the coefficient Q as $a_1^{(2)}/R^{(2)}$, or as $a_1^{(3)}/R^{(3)}$, or as $a_1^{(4)}/R^{(4)}$ [using the values of the curvature $R^{(2)} = 2$, $R^{(3)} = 6$, $R^{(4)} = 12$, where here and below the radius of the sphere is 1]. Of course, the result is the same. We find the coefficients U, V, W from the three linear equations with three unknowns

$$a_2^{(n)} = UR^{(n)^2} + VR^{(n)ik}R_{ik}^{(n)} + WR^{(n)iklm}R_{iklm}^{(n)}$$

(equations for $n = 2, 3, 4$).

In Ref. 4 approximately the same method was used to consider the scalar field without the additional term $-R\phi/6$ in the equation of motion.

The method of Ref. 4 differs from ours in that S^4 is not considered, but instead a relation between U and Q is added ($U = Q^2/2$). For a scalar we readily find

	g	λ
S^2	$2j + 1$	$j(j + 1)$
S^3	$(J + 1)^2$	$J(J + 2)$
S^4	$\frac{1}{6}(I + 1)(I + 2)(2I + 3)$	$I(I + 3)$

The summation for S^3 is particularly simple, the difference terms vanish identically, and $G = G_0 e^\tau$.

Author's Remark[†]

Unfortunately there are some errors in the paper. The consideration of vector and spinor cases is incorrect. This part of the paper is omitted in the present publication. The values of coefficients Q to Y for the scalar equation (without the addition term $-R\phi/6$) are

Q	U	V	W	Y
$\dfrac{1}{6}$	$\dfrac{1}{72}$	$-\dfrac{1}{180}$	$\dfrac{1}{90}$	$\dfrac{1}{30}$

The coefficient Y is determined by considering $G(0)$ with metric having $R \neq \text{const}$:

$$ds^2 = dr^2 + S(r)\, d\phi^2$$

Of greatest importance is the erroneous sign in the formula (33) for the gravitational constant. This error originated in the author's incorrect passage from the euclidean to the pseudoeuclidean case in the formula (30) (the sign in front of the term iQ). After correcting this error, one sees that the increase of $1/G$ has the sign opposite to that of $C_i Q_i$. Thus we arrive at a considerable difficulty with the sign of the gravitational constant in the theory of the zero Lagrangian.

REFERENCES

1. A. D. Sakharov, *Dokl. Akad. Nauk SSSR* 177:70 (1967); *Sov. Phys. Dokl.* 12:1040 (1968), trans. [Paper 14 in this volume].
2. A. D. Sakharov, in *Gravitation and Field Theory* (Collection of Preprints of

[†]Written for this volume—EDS.

the Institute of Applied Mathematics), Moscow (1967); Paper 15 in this volume, trans.
3. L. D. Landau and I. Ya. Pomeranchuk, *Dokl. Akad. Nauk SSSR 102*:489 (1955). E. S. Fradkin, *Dokl. Akad. Nauk SSSR 98*:47 (1954); *100*:897 (1955). Ya. B. Zel'dovich, *ZhETF Pis'ma* 6:1233 (1967).
4. H. P. McKean, Jr. and I. M. Singer, *J. Differ. Geom.* 1:43 (1967).
5. Ya. B. Zel'dovich, *Uspekhi Fiz. Nauk 95*:209 (1968); *Sov. Phys. Uspekhi 11*:381, trans.; oral communication, 1967.
6. D. S. De Witt, *Phys. Rev. 160*:1113 (1967); *162*:1192, 1239 (1967).
7. V. Weisskopf, *Kgl. Danske Veid. Selsk. 14*:1 (1936).
8. I. A. Batalin and E. S. Fradkin, *Preprint FIAN 137*, P. N. Lebedev Physics Institute, Moscow (1967).
9. V. A. Fock, *Izv. Akad. Nauk SSSR, Otd. Mat.-Est. Nauk* 5:51 (1937).
10. J. Schwinger, *Phys. Rev. 82*:664 (1951).

Paper 17

Scalar-Tensor Theory of Gravitation

It is noted that the hypothesis previously advanced by the author, that the Lagrangian of the gravitational field is zero, leads in principle to the impossibility of observing the scalar field of the scalar-tensor gravitation theory, which thus goes over into Einstein's pure-tensor theory. This conclusion is the result of the special form of the dependence of the gravitational constant on the scalar field, $G = \mu^{-2}$ (μ is the scalar field, and the particle masses are proportional to μ).

The scalar-tensor theory of gravitation and its possible experimental consequences have been under discussion in the literature for a number of years [1]. This article calls attention to important consequences that follow for this theory from the hypothesis of the zero gravitational-field Lagrangian (ZL), advanced by the author in 1967 [2]. It follows from the ZL hypothesis that the scalar field is unobservable and is excluded from the theory, which thus goes over into Einstein's usual pure-tensor theory.

If, however, we forgo the ZL hypothesis, then the scalar field becomes manifest in observable effects. But at the same time, it is revealed that it is impossible to satisfy the condition of equivalence (proportionality) of the inertial and gravitational masses. We do not regard a theory with the equivalence principle violated as satisfactory.

In the scalar-tensor theory of gravitation one introduces, besides the tensor gravitational field $g_{ik}(x)$, also a certain scalar field $\mu(x)$. It is assumed that the masses of all bodies depend on the coordinates and the time and are proportional to a common variable scalar field $\mu(x)$. We shall show below that in the general case this condition cannot be

Source: О скалярно-тензорной теории гравитации, Письма в ЖЭТФ 20:189-191 (1974). Reprinted from *JETP Lett.* 20:81–82 (1974), with permission from American Institute of Physics.

satisfied in a noncontradictory manner, but for the time being we confine ourselves to the traditional treatment, which actually is valid only if the gravitational mass defect is neglected. The action for a material point is equal to

$$S_t = - \int m(x)\sqrt{dx^i\, dx^k\, g_{ik}} \qquad m(x) \sim \mu(x) \tag{1}$$

For a classical trajectory we have $\delta S_t = 0$. The trajectory of a point is thus determined by the fields g_{ik} and μ.

The action (1) is invariant against the scale transformation

$$g_{ik}^0 = \mu^2 g_{ik}$$

$$m^0 = \frac{m}{\mu} = \text{const} \tag{2}$$

$$dS^0 = \mu\, dS$$

Here dS is an interval in a gauge that admits of the variable μ field, and dS^0 is the transformed interval.

The transformation (2) was investigated in great detail by Dicke [1].

It turns out that it is just the interval dS^0 which has a direct physical meaning. Let us consider by way of example the measurement of time by an atomic clock that moves along a world line in a field $\mu(x)$. The frequency of the atomic oscillations is $\omega = \Delta E/\hbar$ (ΔE is the difference between the atomic levels). The energy is proportional to the rest mass, i.e., to μ. The frequency ω is proportional to μ. The time interval measured by the atomic clock is therefore $\mu\, dS = dS^0$. This conclusion can be extended directly to include clocks that use other nongravitational phenomena (elastic oscillations, the earth's diurnal rotation, etc.). But the extension of this conclusion to gravitational effects calls for the use of additional information on the dependence of μ on the gravitational constant G. For example, the oscillation frequency of a pendulum is $\omega = \sqrt{GM/R^2 l} \sim G^{1/2} \mu^2$ (M is the mass of the earth and is proportional to μ, R is the radius of the earth and is proportional to μ^{-1}, and the pendulum length is proportional to μ^{-1}). If $G \sim \mu^{-2}$, then all clocks vary in the field μ in the same manner, $\omega \sim \mu$, and the field μ is in fact unobservable and is excluded from the theory by the scale transformation (2). The physical consequences of such a theory agree fully with Einstein's general relativity theory.

If we forgo the condition $G \sim \mu^{-2}$, then the field μ is not eliminated by the transformation (2) and becomes manifest in a number of phenomena (the pendulum oscillation frequency is not proportional to μ, $G^0 = \mu^2 G$ is not a constant, etc.). At the same time, however, the equivalence principle is violated. For example if we take the assumption made in Ref. 1 that G = const and $G^0 \approx \mu^2$, then the gravitational mass defect is proportional not to μ but to μ^3. Such a theory seems very unlikely.

The relation $G \sim \mu^{-2}$ and the possibility of eliminating the scalar field by a scale transformation follows from the zero-Lagrangian hypothesis [2]. It is assumed that in a consistent quantum field theory the Lagrangian function and the total action reduce to the Lagrangian function of matter $\mathcal{L}_m$ and to the action of matter $S_m = \int \sqrt{-g}\, \mathcal{L}_m\, dx$, while the action of the gravitational field [phenomenologically equal to $S_g = \int \sqrt{-g}\,(R/16\pi G)\, dx$] is a polarization effect. The action of matter S_m is invariant under the transformation (2) at $\mathcal{L}_m \sim \mu^4$. The ZL hypothesis extends this invariance to include gravitational phenomena.

Conclusion

A scalar-tensor gravitation theory whose conclusions differ from those of general relativity theory cannot be reconciled with the hypothesis that the gravitational field has a zero Lagrangian.

REFERENCES

1. P. Jordan, *Astr. Nach.* 276:1955 (1948). Schwerkraft, Weltall, 1959. G. R. Tirry, *Compt. Rend.* 226:216 (1948). P. Bergmann, *Ann. Math.* 29:255 (1948). C. Brans and R. H. Dicke, *Phys. Rev.* 124:925 (1961). R. H. Dicke, *Phys. Rev.* 125:2163 (1962).
2. A. D. Sakharov, *Dokl. Akad. Nauk SSSR* 177:70 (1967); *Sov. Phys. Dokl.* 12:1040 (1968), trans. [Paper 14 in this volume]. A. D. Sakharov, Preprint, Institute of Applied Mathematics, Moscow, 1967, *Gravitation and Field Theory*; Paper 15 in this volume, trans.; Paper at Seminar, Phys. Inst. Acad. Sci., June 1970.

Paper 18

The Topological Structure of Elementary Charges and *CPT* Symmetry

The hypothesis on the topological structure of elementary charges was first suggested by Wheeler, who examined the vacuum electric field for an object having the topological structure of a "handle" or "wormhole."

In this article we use the ordinary topological terminology. We will examine, for example, a three-dimensional space from which are cut two spheres. By identifying pairwise the points of one sphere with those of the other, we obtain the wormhole, while each sphere is called the "base." The base can have a more complicated topological structure than a sphere (for example, a torus, a knot with one or another number of petals, etc.). The wormhole for a four-dimensional space-time x, y, z, t is defined in an analogous manner.

In his original model, Wheeler examined the spacelike wormhole and base topologically equivalent to a sphere. In his model, without violating the vacuum Maxwell equations, one base could be considered the source of an electric field and the other, the sink. Wheeler examined two such bases as a possible model for positive and negative elementary charges. The idea of the topological structure of elementary charges was further elaborated in another work of Wheeler [1].

Here, we will examine wormholes of another type, admitting an immediate topological interpretation as elementary charges of various types and signs. The hypothesis set forth below most naturally conforms to those charges not possessing long-range interactions, i.e., to baryonic and to two leptonic charges, and not to electric charge. As is well known, long-range interaction is closely connected to the invariance of the Lagrangian with respect to the gradient of the transformation (see, in

Source: Топологическая структура элементарных зарядов и CPT-симметрия, в сборнике «Проблемы теоретической физики», посвященном памяти И. Е. Тамма, «Наука», Москва, 1972., 243–247. Translated from *Problems of Theoretical Physics*, collection dedicated to the memory of I. E. Tamm, Moscow, 1972, by T. R. Rothman.

this connection, Lee and Yang [2]). We do not discuss here the interpretation of the law of conservation of electric charge. For all other charges we postulate a purely topological origin for the laws of charge conservation, thus avoiding the difficulty mentioned in Ref. 2: the absence of long-range interaction and a compensating (gauge) field.

Figure 1 depicts three-dimensional knots with three and five petals, which we will examine as the bases of the wormhole. We suppose that the base knot with any odd number of petals corresponds to one or another type of charge, while the mirror image of this knot corresponds to the charge of the opposite sign (a and $\bar{a}$, b and $\bar{b}$ in Fig. 1).

Turning to kinematics, we will examine the bases in four-dimensional space-time, x, y, z, t. Let the three-dimensional slices of the base have the structure of a knot (schematically depicted in Fig. 2).

If the 3-slice is a spacelike knot, then the base represents a charge or anticharge. A timelike slice of the knot corresponds to creation or annihilation of charge-anticharge pairs. This corresponds to the well-known Zeisman−Feynman interpretation of pair creation and annihilation as a reversal in the particle's world line.

In the upper half of Fig. 2 is drawn a base corresponding to a vacuum loop: point A represents pair creation; point B, the charge; point C, pair annihilation; point D, the anticharge (the 3-slices of the base are drawn schematically).

We further suppose that the wormhole is defined by the pairwise identification of the two bases, belonging to two mirror-image four-dimensional spaces T and $\bar{T}$. In conformity with the author's hypothesis [3] concerning the CPT symmetry of the universe, we assume that T and $\bar{T}$ are two halves of a single four-dimensional space-time. This assumption is independent and not necessary for the hypothesis set forth in this work but it seems to us a natural one. And so, we consider that to every point on the surface of the base cavity with coordinates x, y, z, t, there corresponds a point on the mirror base with coordinates $x, y, z, -t$. The hypersurface of symmetry $t = 0$ is the cosmologically singular hyper-

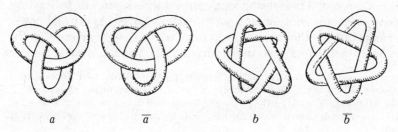

Figure 1

FIELD THEORY AND ELEMENTARY PARTICLES

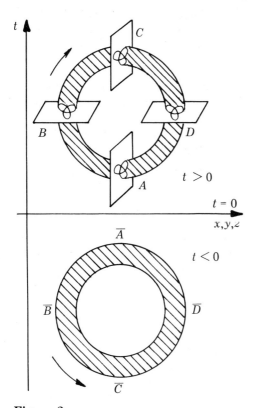

Figure 2

surface of zero extent which does not intersect any base (the initial state of the Friedman universe is neutral in all charges). In Fig. 2 is also schematically depicted the mirror base $\bar{A}, \bar{B}, \bar{C}, \bar{D}$. To the bases in the form of knots we ascribe half-integer spin in conformity with the ideas of Zeisman on the connection between elementary charges and half-integer spin. In quantum theory the state vector is a functional ϕ of the fields ψ_l, defined on the four-dimensional space-time $T + \bar{T}$:

$$\phi = \phi(\psi_l(x, y, z, t))$$

However, by virtue of the hypothesis of CPT symmetry, we can write

$$\phi(T + \bar{T}) - \phi(T)\phi(\bar{T}) = |\phi(T)|^2 = |\phi(\bar{T})|^2 \qquad (1)$$

Therefore, if $\phi(T + \bar{T})$ transforms under three-dimensional rotations as a state vector with odd-integer spin, then $\phi(T)$ and $\phi(\overline{T})$ transform as state vectors with half-integer spin.

When the hypothesis stated in this article is examined seriously, many difficult questions arise:

1. What physical meaning have bases which are not topologically equivalent to knots (those equivalent to a sphere, torus, or to various multiply connected regions with and without torsion)?
2. Which fields and Lagrangian densities must be postulated to determine the dynamics of motion for the charges?
3. How can we carry out the calculation of the slices and mass of the stationary quantum states (particles)?

Only for the second question do we have even a very tentative suggestion.

Let the four-dimensional space-time x, y, z, t, be a Riemann space with torsion. In this case, the connection coefficient Γ^i_{kl} is not symmetric in its lower indices and there exists the gamma tensor,

$$\Gamma^i_{[kl]} = \tfrac{1}{2}(\Gamma^i_{kl} - \Gamma^i_{lk}) \qquad (2)$$

The components of this tensor have the dimensions of stresses of some field. We assume that the Lagrangian density is defined by the invariants of the gamma tensor, Γ_1 of dimension $\mathscr{L}^{-2}$ and Γ_2 of dimension $\mathscr{L}^{-4}$. The invariant Γ_1 is analogous in structure to the well-known electromagnetic invariant $E^2 - H^2$, while Γ_2 is analogous to the invariant $(EH)^2$. Arbuzov and Fillipov have considered the gamma tensor as directly defining the electromagnetic field. Here, we do not take this point of view.

In order for localized stable states to exist, it is essential that there be terms in the Lagrangian which transform differently under similarity transformations. We assume that the Lagrangian density has the form (of course, this is a prior hypothesis)

$$\mathscr{L} = M^2 \Gamma_1 + A\Gamma_1^2 + B\Gamma_2 \qquad (3)$$

Here, we have set $\hbar = c = 1$; the coefficients A and B are ~ 1; and M is a characteristic mass, probably of the order of the Planck mass, $\sim 2 \times 10^{-5}$ g.

We will suppose that the scalar curvature R, which enters into the Hilbert action principle, does not enter into the Lagrangian density. Such an assumption was made by the author in his previous work [4]. There exist several arguments to show that the presence in R of second derivatives of g_{ik} destroys the elegance of the theory. This is especially true if in the Lagrangian density quantities like R enter nonlinearly (which of course is the case in unified field theories). In this instance, even in the expression for the variation of the action, the second derivatives cannot be excluded. In the phenomenological Lagrangian, of course, the quantity R and other analogous invariants enter as usual into the calculation of quantum effects of the vacuum polarization of the gravitational field (cf. Ref. 4).

For quantum theory we assume that in the space x, y, z, t there is defined a series of complex-valued fields $\psi_l(x, y, z, t)$, through the differential forms of which are defined all the tensors and invariants of the theory; for example, we have the metric tensor

$$g_{ik} = \frac{1}{2} \sum_l \left(\frac{\partial \psi_l}{\partial x^i} \frac{\partial \psi_l^*}{\partial x^k} + \frac{\partial \psi_l}{\partial x^k} \frac{\partial \psi_l^*}{\partial x^i} \right) \qquad (4)$$

and the gamma tensor.

The state vector ϕ is a functional of a finite set of the functions ψ_l. The introduction of tensors of the form (4) has the advantage that, under functional integration, one can easily see what remains invariant under coordinate transformations.

The stationary states are superpositions of a state vector that has a given set of elementary charges n_3, n_5, n_7, etc., and an infinite number of state vectors that are distinguished from it by additional charge-anticharge pairs.

In conclusion, we note that the assumption of the construction of elementary particles from a larger number of elementary charges opens new possibilities for explaining the stability of particles. To illustrate these ideas we examine the following hypothetical model. Let leptons with leptonic charge (electron and neutrino) consist of 19 knots with three petals ($n_3 = 19$) and represent them as a regular 20-hedron with one removed facet. Such a structure can have minimal energy if the law of attraction of like knots demands their disposition in the form of a surface. Analogously, for the model of the μ neutrino, we assume a structure in the form of a 12-hedron with one facet removed; moreover, each facet is a knot with five petals ($n_5 = 11$). If now a baryon is represented by a structure consisting of $\bar{n}_3$ knots with three petals and $\bar{n}_5$ knots with five petals (it is possible that $\bar{n}_3$ and $\bar{n}_5$ are divisible by 3; i.e., baryons consist of three quarks), then the disintegration of baryons into leptons is described by the formula

$$209 b(\bar{n}_3, \bar{n}_5) \to 11 \bar{n}_3 l_e(19, 0) + 19 \bar{n}_5 l_\mu(0, 11) \qquad (5)$$

In this manner, the disintegration of baryons in this model is possible only in nuclei with atomic weight not less than 209, and evidently with very small probability.

On the other hand, the possibility of baryon number violation in the nonequilibrium conditions in the initial high-density expanding universe can lead to reactions which produce a surplus of baryons over antibaryons:

$$\text{Entropy} \to 209b + 11\tilde{n}_3 \bar{l}_e + 19\tilde{n}_5 \bar{l}_\mu \tag{6}$$

($\bar{l}_e$ and $\bar{l}_\mu$ are antileptons). This is in agreement with the hypothesis in Ref. 3 of cosmological *CPT* symmetry.

In this model we do not concern ourselves with the unclear questions of the description of electromagnetic effects, strangeness, and charm of the weak and strong interactions.

I am well aware of the great difficulties in the further elaboration of this hypothesis and that it may possibly bear no relationship to reality. But it seems to me that even in this eventuality, the widening of the zone of research will be useful. I hope that this thought would not be alien to the position of Igor Evgenyevich.

REFERENCES

1. J. A. Wheeler, *Superspace*, Report of the 5th Gravitational Conference in Tbilisi, 1968.
2. T. D. Lee and C. N. Yang, *Phys. Rev.* 98:1501 (1955).
3. A. D. Sakharov, *ZhETF Pis'ma* 5:32 (1967); *JETP Letters* 5:24 (1967), trans. [Paper 7 in this volume]. Dubna preprint (1969).
4. A. D. Sakharov, *Dokl.* 177:70 (1967); *Sov. Phys. Dokl.* 12:1040 (1968), trans. [Paper 14 in this volume].

Paper 19

The Quark Structure and Masses of Strongly Interacting Particles

A baryon resonance with the charge 3e, recently reported in the papers of the Goldhabers and co-workers [10] and by Alexander and collaborators [11] can be explained by assuming its quark structure to be four quarks and one antiquark (4q, q̄). It is suggested that such a structure is a "leading" one (i.e., it is contained in the superposition of structures with the maximal weight) for all baryons, including the 56-plet, for which the minimal quark structure is 3q. A unified mass formula (linear in the masses) for mesons and baryons is proposed, based on the assumption that the spin-spin and annihilation interactions involving strange quarks are suppressed (in the same degree for mesons and baryons).

I. INTRODUCTION: A SPATIAL QUARK MODEL FOR BARYONS

The quark model for strongly interacting particles gives a remarkably simple description of their symmetry properties.

It is usually assumed that the lowest 56 baryon states each consists of three quarks ($3q$). The $3q$ state is symmetric in spin-unitary spin variables and consequently antisymmetric in spatial variables since the quarks (q) are fermions. The antisymmetric function of three spatial variables should have a very special character.

The wave functions should vanish for the $3q$ distributed at the corners of an isosceles triangle. The dependence of such a function on the distances between the quarks is studied in Ref. 1. The wave function of particles that attract each other is symmetric with respect to the spatial variables in their lowest energy states. Therefore the antisymmetric

Coauthor: Ya. B. Zel'dovich.
Source: Кварковая структура и массы сильновзаимодействующих частиц, Ядерная физика 4:395-406 (1966) (совместно с Я. Б. Зельдовичем). Reprinted from *J. Nuclear Phys. (USSR)* 4:283–290 (1967), translated by A. M. Bincer, with permission from American Institute of Physics.

character of the wave function of the quarks within the baryons indicates a repulsion which makes it difficult to understand the very stability of the baryons.

The resolution of the difficulty proposed in Ref. 2, namely, the assumption of the existence along with the two-particle repulsive forces, of three particle attractive forces ("six-point function"), seems to us unnatural.

The hypothesis (see, for example, the review Ref. 3[†]) that the baryons contain in addition to quarks a special "center of attraction" also seems to us unnatural; this center may be given a baryonic charge and electric charge $+2e$ or $-e$, with the quarks possessing integer charges and being unstable, in contrast to the version with fractional charges where the p quark is stable and observable. On the other hand, the theory with fractional charges, aside from being more natural, has advantages relating to the description of the magnetic moments and electromagnetic mass differences of the baryons.

Below we develop the hypothesis that baryons consist of four quarks and one antiquark ($\bar{q}$), i.e., that their composition is $4q, \bar{q}$. Since quarks are strongly interacting particles and contribute to the polarization of the vacuum, such an assertion must be viewed as having a conditional character: the physical baryon should be a superposition of the states $(3q)$, $(4q, \bar{q})$, $(5q, 2\bar{q})$, ... However, by analogy to the Tamm–Dancoff method we shall assume that one of these states $(4q, \bar{q})$ appears with the largest amplitude, i.e., is the leading state and determines the basic properties of the particle.

In essence the hypothesis[‡] that the interaction between the q in the conventional scheme $(3q)$ is due to some unitary neutral X meson is practically equivalent to our hypothesis.

Since the X meson polarizes the vacuum and gives the $q, \bar{q}$ pair, it follows that $3q + X = 3q + q\bar{q} = 4q, \bar{q}$. However, our hypothesis differs in that we wish to consider the state $4q, \bar{q}$ as the leading state and all $4q$ (those from "baryonic" $3q$, as well as from the mesonic $q\bar{q}$) will be treated symmetrically. In the $4q, \bar{q}$ model, it is possible to assume a pair-pair ("four-fermion") interaction, namely, attraction between q and $\bar{q}$ and repulsion between q and q. This explains in a natural way why the lowest energy state corresponds to the completely antisymmetric spatial wave function of the $4q$ and the symmetric (with respect to the center of mass of the $4q$) wave function of the $\bar{q}$.

[†] This popular review contains in particular, references to the considerations of L. B. Okun' on the annihilation interaction of mesons.

[‡] Discussed, in particular, in Pais' lecture in Dubna in the autumn of 1965.

Such a function allows a remarkable geometrical interpretation. The lowest four functions in isotropic space are $S_{0,0}$, $P_{1,-1}$, $P_{1,0}$, and $P_{1,1}$, where the subscripts denote l and m_z. The antisymmetric function of four particles is the determinant made out of the foregoing four functions (SP^3). In that state, obviously, $L = 0$. The determinant is unaffected if one goes over from S and P functions to their linear combinations, for which one may choose the four directed σ functions. These functions are introduced in the theory of the valence of carbon; the maximum of the wave function of the system Ψ corresponds to the four particles distributed along the corners of a regular tetrahedron. Under arbitrary rotations of the tetrahedron Ψ remains unchanged (in accordance with $L = 0$). The interchange of any pair of particles, corresponding to the inversion of the tetrahedron or to going over from a right- to a left-handed configuration (for four particles different in spin and unitary spin), changes the sign of Ψ. With this, $|\Psi|^2$ remains unchanged— the right and left configurations are equally probable. We note that $\bar{q}$ forms a symmetric, nodeless wave function (similar to the S wave) in the tetrahedron, thus helping keep the mutually repelling q's together.

The symmetry of the tetrahedron is not necessarily connected with the assumption that the wave functions of the quarks are nonrelativistic. In a relativistic theory of one particle with spin 1/2, along with the total angular momentum of the particle j there is also present an exact quantum number, the parity $P = \pm 1$. To any combination of j and P there corresponds uniquely a fully determined nonrelativistic value of l, for example,

$j = \tfrac{1}{2}$ $P = 1$ $l = 0$

$j = \tfrac{1}{2}$ $P = -1$ $l = 1, \ldots$

Therefore the relativistic generalization of the theory changes quantitatively the distribution of terms in comparison with the solution of the Schrödinger equation (in particular it introduces lS coupling, and splits $P_{1/2}$ and $P_{3/2}$). However, in the relativistic theory, too, it is possible to ascribe to each term a definite l.

The symmetry of the tetrahedron is not connected with the assumption of nonrelativistic motion of the $4q$ in the field of a heavy core or in a self-consistent field, but only with the exact quantum numbers J and P. The tetrahedron is the simplest regular polyhedron in three-dimensional space. This consideration makes the combination $4q$, $\bar{q}$ special among all mq and $n\bar{q}$, and can be considered as the qualitative explanation of the fact that the baryonic charge of the baryon is three times that of the quark.[†]

To the above-described wave function there correspond 756 charge-spin states (756 = 6 × 126, 6 is the number of possible charge-spin states of the antiquark, 126 is the number of different states of four quarks in a symmetric charge-spin state). The intrinsic parity of all these states is the same, namely, $-i$ if the parity of the quark is assumed to be $+i$, i.e., the same as in the $3q$ model.

The $SU(6)$ symmetry corresponds to the annihilation interaction $n\bar{n} \rightleftarrows p\bar{p} \rightleftarrows \lambda\bar{\lambda} \rightleftarrows n\bar{n}$ in the states with spin zero. This singles out the well-known group of 56 states that can be described by the $3q$ model, since in these states $q\bar{q}$ forms an $SU(6)$ singlet. The remaining 700 states correspond to another irreducible representation of the $SU(6)$ group. Griffith [4] has analyzed by symmetry-theory methods all multiplets of this representation, corresponding to different spins ($J = 1/2$, 3/2, 5/2), isospins, and strangenesses. Considered in particular detail, with concrete mass relations, is the case $J = 5/2$. He also cites some earlier papers [5–8].

Let us note that in quark models one obtains, naturally (see Ref. 9), a larger mass for the baryon with strangeness +1 than for the mass of the baryon with strangeness 0, and degeneracy in mass for the four pairs of multiplets for $J = 5/2$ (see Sec. IV).

Originally the $4q, \bar{q}$ model was suggested in connection with ideas about the triply charged resonance N^{+++}, decaying into $p + \pi^+ + \pi^+$ (see Refs. 10 and 11). However, now we propose the hypothesis that all baryons have such a structure (subject to the reservations and limitations described above). This hypothesis may be reasonable even in the case if, in agreement with the considerations given in Ref.10, the state N^{+++} with $M = 1570$ MeV is not a resonance but a dynamical maximum in scattering ("enhancement"). Some of the more important prob-

---

[†] In Ref. 31 the fact that precisely $3q$ (and not $2q$ or $4q, 5q, \ldots$) constitutes a stable system is connected with the dependence of the mass of the system on the number of q: once nq have formed a light system, this system no longer can absorb an additional q just because it is light. The number $n = 3$ itself is not explained, and the hypothesis is altogether not convincing. This paper, however, notes correctly another important circumstance: it is necessary that the interaction between two quarks be described not by a simple potential depending on the distance between them, but one containing Majorana exchange forces. Let us consider different q_1 and q_2. The mutual repulsion of q_1 and q_2 will ensure a large distance between them, but then there remains the possibility of a wave function which is symmetric or antisymmetric with respect to the exchange P_{12}. For arbitrary $u(r_{12})$ the symmetric state is lower; one needs exchange forces in order that, in agreement with what is known about the octet and decuplet, the lower state be totally antisymmetric for any choice of q_1 or q_2.

lems are the clarification of the nature of the resonance N^{+++}, its spin and parity, and the search for other members of the 700-plet, in particular with $S = +1$ and $S = -4$.

At the present time baryon resonances with higher spin and negative parity are not known to be contained in the 756-plet; they must be thought of as states with orbital angular momentum different from zero. (For similar considerations for meson resonances see, for example, Ref. 12, and in the quark model for baryons see Ref. 13.) Along with orbital excitation of individual quarks, the rotation of the tetrahedron as a whole is possible. It is self-evident that in the absence of a rigorous and complete theory the quark models can reflect various properties of the true situation only in part and in a rough form; in particular, the masses of free quarks are unknown, which is of principal importance in view of the lack of experimental data on the existence of free quarks in the mass range up to 6–8 BeV. The simplest application of quark models is a study of regularities that determine hadron masses and the classification of particles; in Sec. II we discuss the masses of the mesons 0 and 1, the three-quark model of the baryon 56-plet, and the mass of the baryons with spin 5/2. The literature on quark models is quite extensive. (Aside from the papers already cited see Refs. 14 to 21.) In many of these papers one encounters features characteristic for the given investigation (for example, the mass formula linear in the meson masses in Refs. 12, 18, and 21, the degeneracy in the mass of the ρ and ω in Refs. 15 and 17, the nonrelativistic approximation for the quarks in Refs. 14 to 16 and in the remaining papers, the mixing of isomultiplets with the same values of J, I, S in many papers, etc.). In the papers that make use of symmetry theory (see for example, Refs. 22 to 24) to explain effects such as the difference in mass of Σ or Λ, one introduces quadratic-in-spin terms which depend on the strangeness. We recall here also some of the older papers based on the Sakata model (for example, Ref. 25, where a number of the results of the unitary symmetry have been anticipated, and also Ref. 26, where meson resonances are predicted and a method is given for their detection from the distribution of the reaction products in invariant masses).

The theory is built on the assumption that the mass of an isolated individual quark is very large; also very large (in magnitude) is the negative energy of attraction of any quark-antiquark pair (see Refs. 1 and 7), more precisely its part which is independent of spin or of the type of quark; in other words, a_0 in Eq. (1) (see below) is a small difference of large quantities.

In Sec. II we consider the mesons 0 and 1. We find (for the meson masses and not for the square of the masses) a mass formula which agrees with experiment no worse than the previously known one. The most important conclusions are as follows: (1) The spin-spin interaction of strange q and $\bar{q}$ is substantially weaker than for nonstrange: the reduction factor is given by $(K^* - K)(\rho - \pi) \approx 0.58$ (the number with the baryon masses taken into account is given below). (2) The $q\bar{q}$ annihilation, as is usually assumed, proceeds only in the state with spin 0. The contribution of the annihilation to the energy is positive. The contribution to the energy of the annihilation channel is of the same order as other terms of the mass formula. Therefore the content of the superpositions which constitute the η and X mesons differs strongly from that given by the $SU(3)$ symmetry.

In Sec. III we consider the octet and decuplet of baryons in the three-quark model. The measured masses are in good agreement with the conclusions, obtained in the study of the mesons, about the weaker spin-spin interaction of strange quarks. The corresponding reduction coefficient is the same for mesons and baryons:

$$1 - \alpha = \frac{K^* - K}{\rho - \pi} = 1 - \frac{3}{2}\frac{\Sigma - \Lambda}{\Delta - N}$$

Also equal are the differences in the masses of the quarks, calculated from the masses of mesons and masses of baryons,

$$b = \Lambda - N = \frac{3K^* + K}{4} - \frac{3\rho + \pi}{4}$$

In Sec. IV we consider some simpler consequences following from the quark model as well as the hypothesis presented in that section in application to other members of the 756-plet (masses for $J = \frac{5}{2}$, etc.).

II. MESONS: THE q, $\bar{q}$ MODEL AND THE MASS FORMULA

For the mesons 0^- and 1^- (aside from the "truly neutral" η and X; see below) we propose the formula

$$m_0 = a_0 + b_0 \Sigma |S_i| + c_0 \mu_1 \mu_2 \tag{1}$$

where $\mu_i = (1 - \alpha S_i)\sigma_i$, and S_i is the strangeness of the i-th particle ($S_i = 0$ for p, n; $S_i = +1$ for $\bar{\lambda}$; $S_i = -1$ for λ).[†] This formula corresponds to the assumption that the strange quark λ is heavier by the amount b_0

[†] We can also write $\alpha/(1 - \alpha) = [(\rho - \pi) - (K^* - K)]/(K^* - K) = \frac{3}{2}(\Sigma - \Lambda)/(\Xi^* - \Xi)$. This relation differs just by a factor $\frac{3}{2}$ from Eq. (1) in Ref. 24, based on $SU(6)$, whose authors apply their formula to the masses squared, and this gives rise to disagreement with experiment.

than the nonstrange p, n quarks or that, by assumption, its interaction compared with them is weaker; the terms a_0 and $b_0 \Sigma |S_i|$ describe a spin-independent contribution to the meson mass. The term $c_0 \mu_1 \cdot \mu_2$ describes the spin-spin interaction, equal to $c_0 \sigma_1 \cdot \sigma_2$ for $p\bar{p}$, $n\bar{n}$, $n\bar{p}$; $c_0(1-\alpha)\sigma_1 \cdot \sigma_2$ for $p\lambda$, $n\lambda$; $c_0(1-\alpha)^2 \sigma_1 \cdot \sigma_2$ for $\lambda\bar{\lambda}$. The optimal choice of constants (with baryon masses taken into account; see below) is $a_0 = 598$ MeV, $b_0 = 180$ MeV, $c_0 = 620$ MeV, and $\alpha = 0.42$. The agreement with the observed masses for the π, K, ρ, K^*, ω, and ϕ mesons is shown in Table 1.

When considering η and X $(0-, I = 0)$, we take into account the annihilation channel by introducing into the Hamiltonian a four-particle interaction of the form

$$H_a = d_0 f_i f_k \tag{2}$$

Here $f_i = 1$ for the pair of ordinary quarks and $f_i = 1 - \beta$ for the pair of strange quarks.

In Eq. (2) we omit for brevity factors depending on the spin, which make H_a vanish for vector mesons [in accordance with the fact that the masses of ρ, ω, and ϕ satisfy the formula (1)]. The complete Hamiltonian will be written in the form of a sum of matrix operators in the representation in which the basis vectors are all possible pairs $q\bar{q}$ (nine pairs):

$$H = H_a + H_b \tag{3}$$

Table 1

Particle	$\Sigma\|S_i\|$	Π_{00}	$\Pi_{0\lambda}$	$\Pi_{\lambda\lambda}$	m_{theor}	m_{exp}
π	0	$-3/4$			133	137
ρ	0	$+1/4$			753	750
K	1		$-3/4$		490	494
K^*	1		$+1/4$		868	890
ω	0	$1/4$			753	780
ϕ	2			$1/4$	1010	1020
N	0	$-3/4$			928	939
Δ	0	$+3/4$			1238	1238
Σ	1	$1/4$	-1		1195	1193
Λ	1	$-3/4$			1108	1115
Ξ	2		-1	$1/4$	1340	1317
Y	1	$1/4$	$1/2$		1375	1385
Ξ^*	2		$1/2$	$1/4$	1520	1530
Ω	3			$3/4$	1675	1675

Constants: $a_0 = 598$ MeV; $a_1 = 1083$ MeV; $b_0 = 180$ MeV; $b_1 = 180$ MeV; $c_0 = 620$ MeV; $c_1 = 207$ MeV; $\alpha_0 = 0.42$; $\alpha_1 = 0.42$.

Here H_b is the "diagonal" operator described by Eq. (1), and H_a is the operator with nine nonzero elements, corresponding to the three truly neutral pairs $n\bar{n}$, $p\bar{p}$, $\lambda\bar{\lambda}$ in the state with spin 0:

$$H_a = d_0 \begin{vmatrix} 1 & 1 & 1-\beta \\ 1 & 1 & 1-\beta \\ 1-\beta & 1-\beta & (1-\beta)^2 \end{vmatrix} \begin{matrix} p\bar{p} \\ n\bar{n} \\ \lambda\bar{\lambda} \end{matrix} \qquad (4)$$
$$\quad\;\; p\bar{p} \quad\;\; n\bar{n} \quad\;\; \lambda\bar{\lambda}$$

Out of the three neutral pairs $p\bar{p}$, $n\bar{n}$, $\lambda\bar{\lambda}$ it is necessary to construct linear combinations which are eigenvectors of $H_b + H_a$. This immediately splits off the combination $(1/\sqrt{2})(p\bar{p} - n\bar{n}) = \pi^0$, for which $H_a\pi = 0$, and which enters into the isotriplet π^0, $\pi^\pm$. Two other combinations give the states η and X.

Knowing H_b and the experimental masses of η and X we find the constants

$$d_0 = 580 \text{ MeV} \qquad \beta = 0.75 \qquad (5)$$

The coefficients in the superposition have here the following values:

$$\eta = 0.54\, p\bar{p} + 0.54\, n\bar{n} - 0.65\, \lambda\bar{\lambda} \qquad (5a)$$
$$X = 0.46\, p\bar{p} + 0.46\, n\bar{n} + 0.76\, \lambda\bar{\lambda}$$

It is interesting to compare the situation for the η and X mesons with the predictions of $SU(3)$ or $SU(6)$ symmetries. In the symmetry theories it is assumed that α, $\beta \ll 1$ and $b \ll d_0$, and that the annihilation channel which is symmetric in the three pairs dominates, so that as a result the singlet is split off:

$$X = \frac{1}{\sqrt{3}}(p\bar{p} + n\bar{n} + \lambda\bar{\lambda}) \qquad (6)$$

The isoscalar term of the $SU(3)$ octet, the η^0 meson, would have in exact $SU(3)$ the composition

$$\eta = \frac{1}{\sqrt{6}} p\bar{p} + \frac{1}{\sqrt{6}} n\bar{n} - \sqrt{\frac{2}{3}} \lambda\bar{\lambda}$$
$$= 0.41 p\bar{p} + 0.41 n\bar{n} - 0.82 \lambda\bar{\lambda}$$

The foregoing composition differs strongly from that expected from $SU(3)$ symmetry.

The Gell-Mann–Okubo mass formulas correspond to diagonal matrix elements of the interaction that violates $SU(3)$, and correspond to calculation of the corrections by using unperturbed wave functions.

In considering the states ω, φ, η, and X we take into account the not inappreciable change in the wave functions, due to mixing. From that point of view the good agreement of the Gell-Mann–Okubo formula for the square of the mass of the η^0 meson is an accident—their formula does not take into account the change in the wave function.

In accordance with our assumptions, $a_0 = 2M_q - U_{12}$. Here U_{12} is a quantity of the type H_b, i.e., it does not transform one particle into another.

All terms in Eq. (1), i.e., a_0, b_0, c_0, and d_0, are quantities of the same order of magnitude and each term is small in comparison with M_q and U_{12}. Therefore the interactions that violate $SU(3)$ change the structure (wave function) of the particles only for the truly neutral pairs $q\bar{q}$.

Finally, let us consider the cardinal problem of the nature of interaction of quarks with one another. It may be described by a four-fermion interaction, or in accordance with the perennial example of electrodynamics, as the interaction through some meson field. Both H_b and H_a may be described with the help of a neutral field.

The form of H_a requires a pseudoscalar or axial meson field M (a remark due to L. B. Okun'). With this, should M have a mass larger than the masses of the given π, η, and X mesons, then it follows from general principles that the annihilation channel would lower the energy of the $SU(3)$ singlet in comparison with the octet members. Thus a definite difficulty arises in the meson description of the quark interaction. We must either turn to the four-fermion quark interaction, in which the sign of the constant can be arbitrary, or else assign unusual properties to the meson M which carries the interaction. It is necessary to assume that the mass of M is small (say, equal to zero), but then we are faced with the difficulty that such particles, without doubt, are not observed in a free state.

In conclusion of this section we note one general result: the interaction of the strange λ quark (spin and annihilation) is smaller than for the p and n quarks.

From this should follow the suppression of reaction channels in which the number of pairs $\lambda\bar{\lambda}$ changes, for example, $\pi^+ + p = K^+ + \Sigma^+$ ($p\bar{n} + ppn = p\bar{\lambda} + pp\lambda$) or $p + \bar{p} = \phi + n\pi$, in comparison with $p + p = \omega + n\pi$. Such a suppression was noted earlier. In Ref. 27 it was assumed that the explanation lies within the symmetry theory.[†] From our point of view what is happening here is precisely the noticeable violation of symmetry, such as the different interaction of the λ quark from the interaction of the p and n quarks.

[†]The same question is discussed in Ref. 25 in the Sakata model.

III. BARYONS: THE $3q$ MODEL AND THE MASS FORMULA

Let us apply the mass formula of the form (1) to baryons (octet and decuplet), regarding them as systems constituted out of $3q$, with a completely symmetric spin–unitary-spin wave function

$$m_1 = a_i + b_1 \Sigma |S_n| + c_1 \Sigma \mu_n \mu_m \qquad (7)$$
$$\mu_n = (1 - \alpha_1 S_n)\sigma_n$$

Here the summation is to be carried out over $n,m = 1, 2, 3$, $n > m$; the notation is analogous to that of Eq. (1). In the decuplet the total spin equals $\frac{3}{2}$, from which it follows that all pairs of spin products $\sigma_i \cdot \sigma_k$ have their maximum value $+\frac{1}{4}$—the same as in the addition of two spins equal to $\frac{1}{2}$ into a triplet.

In the octet the total spin is $\frac{1}{2}$; two identical quarks must, as a consequence of antisymmetry due to the Pauli principle and antisymmetry of the orbital wave function, necessarily be in a symmetric spin state, i.e., in the triplet state the total spin is 1. From this, for example, for $\Sigma^+ = pp\lambda$, it follows that $\overline{\sigma_{p_1}\sigma_{p_2}} = \frac{1}{4}$ and $\overline{\sigma_\lambda \sigma_{p_1}} = \overline{\sigma_\lambda \sigma_{p_2}} = -\frac{1}{2}$. With the help of isotopic rotation we find $\overline{\sigma_\lambda \sigma_p} = \overline{\sigma_\lambda \sigma_n} = -\frac{1}{2}$ for $\Sigma^0 = (pn\lambda)$. On the other hand $\Lambda^0 = pn\lambda$ contains a singlet $[pn]$, hence $\overline{\sigma_p \sigma_n} = -\frac{3}{4}$ and $\overline{\sigma_\lambda \sigma_p} = \overline{\sigma_\lambda \sigma_n} = 0$. If we write

$$\Sigma \mu_n \mu_m = \Pi_{00} + (1 - \alpha)\Pi_{0\lambda} + (1 - \alpha)^2 \Pi_{\lambda\lambda}$$

then the mass can be expressed in terms of the "structure" coefficients Π given in Table 1 [obviously, $\Pi = \Pi_{00} + \Pi_{0\lambda} + \Pi_{\lambda\lambda} = -\frac{3}{4}$ for the octet and $+\frac{3}{4}$ for the decuplet; $J(J + 1) = 3 \times \frac{3}{4} + 2\Pi$]. The mixing of the states Σ^0 and Λ^0 occurs only in the order in which the isotopic invariance is violated, and violation of $SU(3)$ or $SU(6)$ only is insufficient.

We choose the following values for the constants: $a_1 = 1083$, $b_1 = 180$, $c_1 = 207$, and $\alpha_1 = 0.42$. The agreement between calculated and measured masses is given in Table 1.

Let us compare the constants in Eqs. (1) and (7). The difference between a_1 and a_0 is quite natural. The equality $b_1 = b_0$ speaks in favor of interpreting this term as the difference in the masses of λ and the masses of p,n (and not the difference in the main interaction U)—it is precisely in this case that we would obtain an additive constant in $\Sigma|S|$, independent of the orbital wave function.

The difference $c_1 \neq c_0$ is natural, since the interaction between spins depends on the orbital wave functions—roughly speaking, on the average distance between quarks, which may be substantially different in a meson and a baryon; in the meson q and $\bar{q}$ are in an S state with respect to each other, i.e., they are "in contact":

$$\Psi(r_{12} = 0) \neq 0$$

In the baryon the antisymmetry of the orbital wave function with respect to interchange of quarks means that each pair of quarks is in a P state with respect to each other and $\Psi(r_{12} = 0) = 0$. It is therefore quite natural that $c_1 = 207$ is one-third as large as $c_0 = 620$ MeV.

It is necessary in particular to emphasize that mesons and baryons can be described satisfactorily with expressions containing identical α, $\alpha_0 = \alpha_1$. This constitutes a real confirmation of the proposed scheme. The total number of independent constants describing eight baryons and six "nonannihilating" mesons is equal to 6.

For baryons Eq. (7) for each multiplet separately does not differ much in structure from the Gell-Mann–Okubo formula. This is natural, since the "structure" of baryons is uniquely determined; the violations of $SU(3)$ do not change the structure (in distinction from the situation for ϕ, ω, or η, X mesons). We would obtain exactly the Gell-Mann–Okubo formula if we replaced $(1 - \alpha)^2$ for the $\lambda\lambda$ interaction by $1 - 2\alpha$, i.e., in the limit when $\alpha \ll 1$.

IV. CLASSIFICATION OF STATES AND MASSES IN THE $4q$, $\bar{q}$ MODEL FOR $L = 0$

The four quarks constitute 126 states with symmetric spin–unitary-spin function (6 states of the type $AAAA$, 30 of the type $AAAB$, 15 of the type $AABB$, 60 of the type $AABC$, 15 of the type $ABCD$); one antiquark has six states A, altogether: $6 \times 126 = 756$ states. To these states correspond the representations 56 and 700 of $SU(6)$; the contents of these two representations classified according to J spin and $SU(3)$ are well known (see Refs. 4, 28, and 29):

$$56 = 10(4) + 8(2)$$
$$700 = 35(6 + 4) + 27(4 + 2) + 10(6 + 4 + 2)$$
$$+ 10^*(2) + 8(4 + 2)$$

The decomposition of the 756 states into the representations 700 and 56 [the latter contains the $SU(3)$ octet with spin $J = \frac{1}{2}$ and $SU(3)$ decuplet with $J = \frac{3}{2}$] occurs only in the presence of $SU(6)$ symmetry; if $SU(6)$ symmetry is violated, mixing of the 56-plet with certain multiplets of the 700-plet takes place, but it is still possible to assume that the 56 states are different in mass, and it is they (see Sec. I) which constitute the lowest states.

The difficulty of this scheme consists, perhaps, in the fact that for mesons the annihilation interaction increases the mass of η and X mesons, i.e., it corresponds to repulsion of q and $\bar{q}$. For baryons our hypothesis requires that at least one octet and one decuplet with annihilation interactions should lie lower than the same states for which there is no annihilation interaction. The verification of the compatibility of this hypothesis with other properties of hadrons will become possible only when additional information on baryon resonances is accumulated.

In classification of the states we suppose that isotopic spin constitutes a good quantum number but that the unitary spin may not have a definite value, in the same sense as is true for mesons (ϕ and ω, η and X in Sec. II).

We can first construct all $SU(3)$ representations corresponding to the $4q$, $\bar{q}$ model with the totally symmetric spin–unitary-spin function of the $4q$. We assume here the occurrence of mixing of multiplets belonging to different $SU(3)$ and even different $SU(6)$ representations with the same spatial and isotopic spin. As a consequence the predictions made in Ref. 9 on the basis of the Gell-Mann–Okubo formula are violated.

In Table 2 we give all the isomultiplets for different values of strangeness S (in rows) and mechanical spin J (in columns). In each entry are given the values of I, and the exponent gives the multiplicity of the corresponding multiplets, if this multiplicity exceeds 1.

The masses of baryons which do not contain the same quark and antiquark, as well as masses of baryons obtained from those by isotopic (but not unitary) rotations, should be describable by simple formulas of the type (1) and (7); the masses of the remaining baryons should be describable by equations of the type (4); however, the numerical values of the constants that enter into these formulas and equations have not yet been determined.

Table 2

S	$J = 5/2$	$J = 3/2$	$J = 1/2$
+1	2	2, 1	1, 0
0	$(5/2)$, $(3/2)^2$	$(5/2)$, $(3/2)^4$, $(1/2)^2$	$(3/2)^2$, $(1/2)^4$
−1	2, 1^2	2^2, 1^5, 0^2	2, 1^5, 0^2
−2	$(3/2)$, $(1/2)^2$	$(3/2)^2$, $(1/2)^5$	$(3/2)^2$, $(1/2)^4$
−3	1, 0^2	1^2, 0^3	1, 0
−4	$1/2$	$1/2$	—

It is most desirable to have additional information on baryons and resonances that enter the 700-plet.

Of special interest is the clarification of the spin and parity of the resonance $p\pi^+\pi^+$ with the mass 1570 MeV [10,11]. According to our hypothesis, the parity should coincide with the parity of the 56-plet, and the J spin should be equal to $\frac{3}{2}$ or $\frac{5}{2}$. Indeed, the quark content of $p\pi^+\pi^+$ is $ppppn$, four identical quarks with parallel spins having total spin $J = 2$, and with the addition of the antiquark $J = 2 \pm \frac{1}{2}$ (see Table 2).

The c.m.s. angular dependence of the resonance N^{+++} obtained in Ref. 11 does not contradict the formula $\sigma \sim \cos^2 \theta$. If this is the case, then the reaction

$$p + p \to \Delta^- + N^{+++}$$

proceeds in a p wave in initial and final states and the parity of N^{+++} (taking into account the positive parity of Δ^-) is indeed positive. The absence of a noticeable S wave excludes here the value $\frac{3}{2}$ for the spin of N, i.e., the spin of N^{+++} should equal $\frac{5}{2}$. Of course all these conclusions need to be verified.

In the Appendix we consider the question of the relation of the channels in reactions for the production of resonances with mechanical spin $J = \frac{5}{2}$. These considerations are apparently likewise not in contradiction with the conclusion $J = \frac{5}{2}$ for the 1570 MeV resonance.

Is a mass distribution possible such that the particle with $I = \frac{5}{2}$, $J = \frac{5}{2}$ has a smaller mass than the particle with $I = \frac{5}{2}$, $J = \frac{3}{2}$? At first sight, this seems strange, but one should recall (see for example Ref. 30), that the spin-spin interaction realized by pseudoscalar fields has different signs for different distances between interacting fermions. In a nonrelativistic approximation the interaction potential is given by

$$U = \text{const} \times (\boldsymbol{\sigma}_1 \boldsymbol{\sigma}_2) \left[\frac{\varkappa^2 \exp(-\varkappa r)}{4\pi r} - \delta(\mathbf{r}) \right]$$

We may suppose that the N^{+++} is such a dense structure that in it, in contrast to mesons, the sign of the spin-spin interaction is determined by the contact term (which in fact is "smeared out" over a region determined by the mass of the quark).

The formula of type (7) for the $4q, \bar{q}$ model has the form

$$m = a_2 + b_2 \Sigma |S_i| + c_2 \Sigma \boldsymbol{\mu}_n \boldsymbol{\mu}_m + c_3 \boldsymbol{\mu}_0 \Sigma \boldsymbol{\mu}_n + H_a \tag{8}$$

Here H_a is the annihilation operator; $\boldsymbol{\mu}_0 = (1 - \alpha_2 S_0)\boldsymbol{\sigma}_0$ refers to the antiquark, $\boldsymbol{\mu}_1, \ldots, \boldsymbol{\mu}_4$ are defined as in Eq. (7), and the indices $n, m = 1, \ldots, 4$ refer to the quarks. The case $J = \frac{5}{2}$ is particularly simple

($\sigma_0\sigma_n = \sigma_m\sigma_n = \frac{1}{4}$; the annihilation term H_a equals 0 in analogy with the 1^- meson—in both cases the quark spins are parallel). The case $J = \frac{5}{2}$ is considered in Ref. 4 within the framework of $SU(6)$ symmetry. However, considerations in the quark model are considerably simpler. It follows from Eq. (8) that the masses are determined only by the strangeness of the antiquark and the number of strange quarks and consequently the masses are the same for the four pairs of multiplets ($I = \frac{5}{2}$ and one of the two multiplets with $I = \frac{3}{2}$ for $S = 0$, and analogously for $S = -1, -2, -3$). This situation is fully analogous to that for the ρ and ω mesons.

Using the parameters $b = 180$ MeV and $\alpha = 0.42$ from Table 1 and supposing that the resonance $p\pi^+\pi^+$ obtained in Refs. 10 and 11 has spin $J = \frac{5}{2}$, we find the remaining nine masses (Table 3). The constants c_2 and c_3 are unknown; it should be presumed that $c_3 < 0$ and $c_2 \approx 200$ MeV. We note (in addition to what is given in the table), that the mass of the particle with charge $+3e$, zero strangeness, and J spin equal to $\frac{3}{2}$, amounts to $1570 - 2.5\,c_3$.

The observation of some of the particles from the 756-plet is made difficult by the high unitary spin; for example, in the scattering $K^+ + p$ it is not possible to have the resonance corresponding to $J = 5/2$ since that resonance has isotopic spin $I = 2$.

CONCLUSIONS

The hypothesis on the $4q, \bar{q}$ structure of baryons explains within the framework of the quark model the stability of baryons, and makes possible the prediction of a number of properties of the baryonic 756-plet. The mass formula for both mesons and baryons can be constructed in the same fashion (linear in masses), if it is assumed that the spin-spin and annihilation interactions are reduced for strange quarks.

Table 3

I	S	m	I	S	m
		$\bar{q} = \bar{n}$ or $\bar{p}$			$\bar{q} = \bar{\lambda}$
5/2, 3/2	0	1570	2	+1	$1750 - 0.42\,c_3$
2, 1	−1	$1750 - 0.32\,c_2 - 0.105\,c_3$	3/2	0	$1930 - 0.32\,c_2 - 0.48\,c_3$
3/2, 1/2	−2	$1930 - 0.585\,c_2 - 0.21\,c_3$	1	−1	$2110 - 0.585\,c_2 - 0.541\,c_3$
1, 0	−3	$2110 - 0.81\,c_2 - 0.315\,c_3$	1/2	−2	$2290 - 0.81\,c_2 - 0.602\,c_3$
1/2	−4	$2290 - 1.0\,c_2 - 0.42\,c_3$	0	−3	$2470 - 1.0\,c_2 - 0.637\,c_3$

APPENDIX

Table 4, column a, lists the theoretical [11] relative yields for the seven channels of the reaction $p + p$ with the production of the resonance $I = \frac{5}{2}$ (decaying into three particles). The ratio of the yields of the two first reactions according to the table amounts to 0.16. Experimentally, one finds 0.6 ± 0.35 (see Ref. 11). If the J spin of the 1570 MeV resonance is equal to $\frac{5}{2}$, then according to Sec. IV the same mass is possessed by the particles from the isomultiplets with $J = \frac{5}{2}$, $I = \frac{3}{2}$. For all reactions except the first, it is necessary to take into account the superposition of the amplitudes with $I = \frac{5}{2}$ and $I = \frac{3}{2}$:

$$\Psi = a\Psi_{5/2} + b\Psi_{3/2}$$

We have determined the ratio of the coefficients a and b from the quark model by looking at, for example, the quark content of the component of the amplitude Ψ, corresponding to the following quantum numbers: Initial state $L = 1$, $J = 1$, $I = 1$, m (projection of J spin) = 1, μ (projection of I spin) = 1; final state $L = 1$; for the two-particle state, $I_2 = J_2 = \frac{3}{2}$, $m_2 = -\frac{3}{2}$, $\mu_2 = +\frac{3}{2}$; the three-particle state, $J_1 = \frac{5}{2}$, $m_1 = \frac{5}{2}$, $\mu_1 = -\frac{1}{2}$. The superposition is for $I_1 = \frac{5}{2}$ and $I_1 = \frac{3}{2}$.

Taking into account the conservation of the quark spin, we find that the three-particle resonance should have the content $ppnn\bar{p}$ (without any admixture of $pnnn\bar{n}$), which occurs when $b/a = \frac{1}{3}$. Using this ratio we have calculated the relative yields of the seven reactions in the last column of Table 4 $(a + b)$.

Table 4

Channels	a	$a + b$
$N\pi^- + p\pi^+\pi^+$	0.5	0.5
$p\pi^- + p\pi^+\pi^0$	0.08	0.109
$p\pi^- + N\pi^+\pi^+$	0.02	0.002
$p\pi^0 + p\pi^+\pi^-$	0.02	0.042
$N\pi^+ + p\pi^+\pi^-$	0.01	0.021
$p\pi^+ + p\pi^0\pi^-$	0.02	0.08
$p\pi^+ + N\pi^+\pi^-$	0.01	0.001

Note Added in Proof (June 20, 1966)

In a discussion in the Summer School in Balaton (Hungary), R. Sokoloff (Berkeley, U.S.A.) developed the hypothesis on the additivity of total cross sections at high energies. The comparison of the cross section for meson-nucleon scattering with the nucleon-nucleon and antinucleon-nucleon cross sections speaks in favor of the $3q$ baryon model and against the $4q, \bar{q}$ model. On the other hand, these data are in agreement with the conclusion made in Secs. II and III on the weaker interactions for the strange quark.

REFERENCES

1. G. Morpurgo and C. Becchi, *Phys. Lett. 17*:352 (1965).
2. T. K. Kuo and L. A. Radicati, *Phys. Rev. 139*:B746 (1965).
3. Ya. B. Zel'dovich, *UFN 86*:303 (1965); *Sov. Phys. Uspekhi 8*:489 (1965), trans.
4. R. Griffith, *Phys. Rev. 139*:B667 (1965).
5. R. Dashen and D. Sharp, *Phys. Rev. 138*:B223 (1965).
6. E. Abers, L. Balázs, and Y. Hara, *Phys. Rev. 136*:B1382 (1964).
7. R. Dashen and S. Frautschi, *Phys. Rev. 137*:B1318 (1965).
8. H. Harari and H. Lipkin, *Phys. Rev. Lett. 13*:345 (1964).
9. Ya. B. Zel'dovich, *ZhETF Pis'ma 2*:340 (1965); *JETP Lett. 2*:216 (1965), trans.
10. G. Goldhaber, S. Goldhaber, T. O'Holloran, and B. C. Shew, *Report of Proc. of 12th Inter. Conf. on High Energy Physics*, Dubna (1964).
11. G. Alexander, O. Benary, B. Reuter, A. Shapira, E. Simopoulou, and G. Yekutieli, *Phys. Rev. Lett. 15*:207 (1965).
12. V. V. Anisovich, *ZhETF Pis'ma 2*:554 (1965); *JETP Lett. 2*:344 (1965), trans.
13. R. D. Hill, *Nuovo cimento 39*:1197 (1965).
14. M. Gell-Mann, *Phys. Lett. 8*:214 (1964).
15. G. Zweig, Preprint 8182. TH-401, 8419/TH-412, CERN (1964).
16. N. N. Bogolyubov, B. V. Struminskiĭ, and A. N. Tavkhelidze, *Preprint D-1968*, Dubna (1965).
17. F. Gursey, T. D. Lee, and N. Nauenberg, *Phys. Rev. 135*:B467 (1964).
18. F. Duimio and A. Scotti, *Phys. Rev. Lett. 14*:926 (1965).
19. P. G. O. Freund and B. W. Lee, *Phys. Rev. Lett. 13*:592 (1964).
20. P. G. O. Freund, *Nuovo cimento 39*:769 (1965).
21. Ya. I. Azimov, V. V. Anisovich, A. A. Ansel'm, G. S. Danilov, and I. G. Dyatlov, *Yad. Fiz. 2*:583 (1965); *Sov. Phys. JNP 2*:417 (1966), trans.
22. T. Kuo and T. Yao, *Phys. Rev. Lett. 13*:415 (1964).
23. B. Beg and V. Singh, *Phys. Rev. Lett. 13*:418 (1964).
24. H. Harari and H. Lipkin, *Phys. Rev. Lett. 14*:570 (1965).
25. L. B. Okun', *JETP 34*:468 (1958); *Sov. Phys. JETP 7*:322 (1958), trans.

26. Ya. B. Zel'dovich, *JETP* *34*:1644 (1958); *Sov. Phys. JETP* *7*:1130 (1958), trans.
27. H. Lipkin, *Phys. Rev. Lett.* *13*:590 (1964).
28. J. C. Carter, J. J. Coyne, and S. Meshkov, *Phys. Rev. Lett.* *14*:523 (1965).
29. C. L. Cook, G. Murtaza, *Nuovo cimento* *39*:531 (1965).
30. H. A. Bethe and F. de Hoffman, *Meson and Fields*, Vol. II, Row, Pederson & Co. (1955).
31. G. Morpurgo, *Phys. Lett.* *20*:684 (1966).

Paper 20

Mass Formula for Mesons and Baryons with Allowance for Charm

The mass formula previously proposed by Ya. B. Zel'dovich and the author for mesons and baryons is applicable to particles that have charm.

The recent discovery of a vector particle ψ with mass 3105 GeV and with anomalously small hadronic-decay width, and its interpretation as consisting of charmed quarks [1], make particularly essential an exhaustive analysis of the consequences of the hypothesis that a fourth quark exists. In this article we attempt to estimate the masses of mesons and baryons having charm with the aid of the mass formulas previously proposed by Ya. B. Zel'dovich and the author [2]. This formula is based on the "naive" model of nonrelativistic quarks and is written in unified form for mesons and baryons, namely, additively in the masses:

$$M = \delta + \Sigma m_q + b\Sigma\xi_i\,\xi_k(\boldsymbol{\sigma}_i\boldsymbol{\sigma}_k)$$
$$= a + \Sigma(m_\lambda - m_0)|s| + \Sigma(m_\chi - m_0)|c| + b\Sigma\xi_i\,\xi_k(\boldsymbol{\sigma}_i\boldsymbol{\sigma}_k) \qquad (1)$$

Here s is the strangeness; c is an analogous new additive quantum number, which we call in accordance with tradition "charm" and which is equal to the difference between the number of charmed quarks and antiquarks χ and $\bar{\chi}$; m_q is the mass of the quark, $m_q = m_0$ for the "ordinary" quarks p and n (O quarks); δ, b, and a are constants which are different for the meson and baryon. The last term describes the spin-spin interaction of the quarks; $\boldsymbol{\sigma}_i$ is the spin of the ith quark, and ξ_i is the coefficient of attenuation of the spin-spin interaction, $\xi_\chi < \xi_\lambda < \xi_0 = 1$.

It turns out that the constants ξ_λ and the difference $m - m_0$ are

Source: Массовая формула для мезонов и барионов с учетом шарма, Письма в ЖЭТФ 21:554–557 **(1975)**. Reprinted from *JETP Lett.* 21:258–259 (1975), translated by Clark S. Robinson, with permission from American Institute of Physics.

approximately the same for baryons and mesons. For mesons, $\xi_\lambda = (K^* - K)/(\rho - \pi) = 0.645$ and

$$m_\lambda - m_0 = \tfrac{1}{4}(3K^* + K) - \tfrac{1}{4}(3\rho + \pi) = 194 \text{ MeV}$$

For baryons we have

$$\xi_\lambda = 1 - \frac{3}{2}\frac{\Sigma - \Lambda}{\Delta - N} = 0.61$$

$$m_\lambda - m_0 = \Lambda - N = 176 \text{ MeV}$$

These coincidences may possibly indicate a physical meaning in formula (1).

To determine the constants pertaining to the charmed quark, all we know for the time being is the mass of ψ. We assume this particle to consist of $\chi\bar\chi$, neglecting the mixing with the pairs $\lambda\bar\lambda$ and $O\bar O$. We make an additional assumption that connects the constants ξ with the masses of the quarks:

i.e.,
$$\xi_\lambda = \frac{m_0}{m_\lambda} \qquad \xi_\chi = \frac{m_0}{m_\chi} \tag{2}$$

$$\xi_\chi^{-1} = 1 + \frac{m_\chi - m_0}{m_\lambda - m_0}(\xi_\lambda^{-1} - 1)$$

The constants entering into (1) are determined by the same token in a sufficiently unambiguous manner.

For numerical estimates, we assume also that the differences $m_\chi - m_0$ are different for mesons and baryons and are in the same ratio as the differences $m_\lambda - m_0$. We ultimately get the following system of constants:

Table 1

	ϕ	Σ^*	Ξ	Ξ^*	Ω
Formula	1049	1375.5	1337	1520	1672
Experiment	1020	1385	1317	1530	1675

Table 2

J	$O\bar\chi, \bar O\chi$	$\lambda\bar\chi, \bar\lambda\chi$	$\chi\bar\chi$
0	1748.5	1977.5	3076
1	1880.5	2062.5	–

For mesons:
$a = 597$ MeV
$b = 613$ MeV
$m_\lambda - m_0 = 194$ MeV
$m_\chi - m_0 = 1250.5$ MeV
$\xi_\lambda = 0.645$
$\xi_\chi = 0.216$

For baryons:
$a = 1088.5$ MeV
$b = 200.0$ MeV
$m_\lambda - m_0 = 176$ MeV
$m_\chi - m_0 = 1140$ MeV
$\xi_\lambda = 0.61$
$\xi_\chi = 0.195$

We shall explain the calculation of the third term of formula (1), using as an example baryons containing the three different quarks O, λ, and χ. The operator

$$H_{\sigma\sigma} = A(\sigma_1\sigma_2) + B(\sigma_2\sigma_3) + C(\sigma_3\sigma_1)$$

(where $A = b\xi_1\xi_2$, etc.) has eigenvalues

$$E_1 = \frac{A+B+C}{4} \quad (\text{spin } \tfrac{3}{2})$$

$$E_{2,3} = -\frac{A+B+C}{4} \pm \sqrt{A^2 + B^2 + C^2 - AB - BC - CA}$$

$(\text{spin } \tfrac{1}{2})$

The eigenvalues for spin $\tfrac{1}{2}$ were obtained from the two-dimensional secular equation, which can be easily set up by recognizing that the three operators $(\sigma_1\sigma_2)$, $(\sigma_2\sigma_3)$, and $(\sigma_3\sigma_1)$ have the same eigenvalues and are obtained from one another by rotation through 120° in a two-dimensional plane.

The three eigenvalues correspond to three particles Ξ'_c, Ξ''_c, and Ξ^*_c (isodoublets) in Table 3. The remaining cases are even simpler.

The masses of the mesons π, ρ, K, K^*, and ψ and of the baryons N, Δ, Σ, and Λ, with the assumed system of constants, are satisfied identically. For the meson ϕ and the baryons Σ^*, Ξ, Ξ^*, and Ω, formula (1) yields mass values which are very close to the experimental ones (Table 1).

Table 3

J	$OO\chi$		$O\lambda\chi$			$\lambda\lambda\chi$	$O\chi\chi$	$\lambda\chi\chi$	$\chi\chi\chi$
	Σ_c, Σ_c^*	Λ_c	Ξ'_c	Ξ^*_c	Ξ''_c	Ω_c	N_{cc}	Ω_{cc}	Ω_{ccc}
1/2	2239.5	2078	2313		2404	2575	3331.5	3522.5	–
3/2	2298	–		2450.5		2611	3390	3559.0	4514

In Table 2 are gathered the predictions of formula (1) for mesons containing one or two charmed quarks, and in Table 3 are gathered the formulas for baryons containing 1, 2, or 3 charmed quarks.

We note that the mass of the pseudoscalar meson $O\bar{\chi}$ (which we shall call K_c) predicted by the linear mass formula (1) is such that the decay $\psi' \to K_c + \bar{K}_c$ is possible. Here ψ' is a second vector meson with an anomalously small hadron-decay width, the mass of which is 3.7 GeV. No such decays have been observed at the present time. The predicted mass of the pseudoscalar meson $\chi\bar{\chi}$ was calculated without allowance for the mixing with χ and η.

Thus, the masses of five mesons and 13 baryons were predicted with the aid of the linear formula.

I am grateful to the members of the theoretical division of the Physics Institute of the Academy of Sciences for a discussion of this work.

REFERENCES

1. Ref. TH 1964-CERN, December 6, 1974.
2. Ya. B. Zel'dovich and A. D. Sakharov, *Yad. Fiz.* 4:395 (1966); *Sov. J. Nuclear Phys.* 4:283 (1967), trans. [Paper 19 in this volume].

Paper 21

Mass Formula for Mesons and Baryons

The number of independent parameters in the semiempirical formula for meson and baryon masses proposed by Ya. B. Zel'dovich and the author is reduced by applying chromodynamics considerations. The results are compared with experiment. A summary of the new predictions is given.

Ya. B. Zel'dovich and the author proposed a semiempirical formula which gives a unified description of the masses of the mesons and baryons with the wave function of the quarks in the lowest s state (Ref. 1).[†] For mesons we shall write the formula with the minimum number of independent parameters as

$$M = \delta_M + m_1 + m_2 + b\frac{m_0^2}{m_1 m_2}\sigma_1\sigma_2 \tag{1}$$

and for baryons,

$$M = \delta_B + m_1 + m_2 + m_3 + \frac{b}{3}\left(\frac{m_0^2}{m_1 m_2}\sigma_1\sigma_2 + \frac{m_0^2}{m_2 m_3}\sigma_2\sigma_3 + \frac{m_0^2}{m_3 m_1}\sigma_3\sigma_1\right) \tag{2}$$

Here δ_M, δ_B, and b are parameters with the dimensions of mass and m_i are the masses of the quarks. Below we use the following notation: m_o is the mass of the ordinary quark u or d; m_s and m_c are the masses of the strange and charmed quarks. Altogether there are six parameters.

Source: Массовая формула для мезонов и барионов, ЖЭТФ 78:2112-2115 (1980). Reprinted from *Sov. Phys. JETP* 51:1059–1060 (1980), translated by Clark S. Robinson, with permission from American Institute of Physics.

[†] In Ref. 1 in addition to a linear mass formula we discuss also the model of construction of a baryon from four quarks and an antiquark, which is not of interest here.

The last term in Eqs. (1) and (2) is the spin-spin interaction of the quarks; $\boldsymbol{\sigma}_i \cdot \boldsymbol{\sigma}_j$ is the scalar product of the quark spin vectors; $\boldsymbol{\sigma}_1 \cdot \boldsymbol{\sigma}_2 = -\frac{3}{4}, +\frac{1}{4}$, respectively, for pseudoscalar and vector mesons. With total baryon spin $J = \frac{3}{2}$ we have

$$\boldsymbol{\sigma}_1\boldsymbol{\sigma}_2 = \boldsymbol{\sigma}_2\boldsymbol{\sigma}_3 = \boldsymbol{\sigma}_3\boldsymbol{\sigma}_1 = \tfrac{1}{4}$$

With total baryon spin $J = \frac{1}{2}$, if the baryon contains two identical quarks q_2 and q_3 we have

$$\boldsymbol{\sigma}_2\boldsymbol{\sigma}_3 = \tfrac{1}{4} \qquad \boldsymbol{\sigma}_1\boldsymbol{\sigma}_2 = \boldsymbol{\sigma}_1\boldsymbol{\sigma}_3 = -\tfrac{1}{2}$$

However, if the baryon contains three different quarks, then the eigenvalues of the spin-spin interaction operator

$$H_{\sigma\sigma} = A\boldsymbol{\sigma}_1\boldsymbol{\sigma}_2 + B\boldsymbol{\sigma}_2\boldsymbol{\sigma}_3 + C\boldsymbol{\sigma}_3\boldsymbol{\sigma}_1$$

are

$$h_{1/2} = -\tfrac{1}{4}(A + B + C) \pm 2^{-3/2}[(A - B)^2 + (B - C)^2 + (C - A)^2]^{1/2}$$
$$h_{3/2} = +\tfrac{1}{4}(A + B + C) \tag{3}$$

These eigenvalues can be found by direct solution of the secular equation of eighth order. We previously [2] pointed out a simple procedure for separating the second-order secular equation and for finding $h_{1/2}$. We shall consider a two-dimensional space corresponding to spin $\frac{1}{2}$, and a definite projection. The two-dimensional operators $\boldsymbol{\sigma}_i \cdot \boldsymbol{\sigma}_j$ have eigenvalues $-\frac{1}{4} \pm \frac{1}{2}$ and are obtained from each other by rotation of the basis by 120° (from symmetry considerations). If one of them is diagonal in a certain basis,

$$\alpha = -\frac{1}{4} + \frac{1}{2}\begin{pmatrix} 1 & 0 \\ 0 & -1 \end{pmatrix}$$

then the other two in the same basis are

$$\beta = -\frac{1}{4} + \frac{1}{2}\begin{pmatrix} -\frac{1}{2} & \frac{\sqrt{3}}{2} \\ \frac{\sqrt{3}}{2} & \frac{1}{2} \end{pmatrix} \qquad \gamma = -\frac{1}{4} + \frac{1}{2}\begin{pmatrix} -\frac{1}{2} & -\frac{\sqrt{3}}{2} \\ -\frac{\sqrt{3}}{2} & \frac{1}{2} \end{pmatrix}$$

The quantities $h_{1/2}$ are the eigenvalues of the matrix

$$H'_{\sigma\sigma} = \alpha A + \beta B + \gamma C$$

Equation (3) describes the masses of the isodoublet baryons of composition (o, c, s), where o is an ordinary quark, and also (for $C = B$) the masses of the doublets Σ, Λ and Σ_c, Λ_c. We note that in Ref. 2 in the corresponding formula we erroneously omitted the factor $\frac{1}{2}$ in front of the radical, but the numerical values have been calculated correctly.

The spin-spin interaction in Eqs. (1) and (2) are interpreted as the interaction of magneticlike gluon moments which are proportional to $g/2m$, by analogy with the ordinary Dirac magnetic moment; g is the interaction constant. Of course, the large value of the radiation and vacuum corrections makes this interpretation somewhat arbitrary (compare Ref. 3). The empirical factor $\frac{1}{3}$ in Eq. (2) can be explained in the following way. The spin-spin interaction is proportional to $(g_1 g_2)/V$, where V is the effective volume of the hadron and $(g_1 g_2)$ is the scalar part of the color charge vectors in two-dimensional charge space. We assume that the effective volumes of the baryons and meson are related as 3:2 in correspondence with the number of quarks. In the case of a meson the scalar product of the charge vectors of the quark and antiquark is equal to $-g^2$. The angles between the charge vectors of the quarks of three different colors entering into the composition of the baryon in charge space are equal to 120°. The scalar products are equal to $g^2 \cos 120° = -g^2/2$. Thus, the scalar products of the charges for the baryon and meson are related as $\frac{1}{2}$:1. Collecting the factors, we have $\frac{2}{3} \times \frac{1}{2} = \frac{1}{3}$.

The relations connecting the mass differences of the baryons and mesons follow from Eqs. (1) and (2). They are adequately satisfied experimentally (see Refs. 4 to 6):

$$\frac{\Delta - N}{\rho - \pi} = \frac{1}{2} \quad \text{(experimentally 0.46)} \tag{4a}$$

$$\frac{\Sigma^* - \Sigma}{K^* - K} = \frac{1}{2} \quad \text{(experimentally 0.48)} \tag{4b}$$

$$\frac{K^* - K}{\rho - \pi} + \frac{3}{2} \frac{\Sigma - \Lambda}{\Delta - N} = 1 \quad \text{(experimentally 1.02)} \tag{4c}$$

$$\frac{D^* - D}{\rho - \pi} + \frac{3}{2} \frac{\Sigma_c - \Lambda_c}{\Delta - N} = 1 \quad \text{(experimentally 1.08)} \tag{4d}$$

$$m_s - m_o = \tfrac{1}{4}(3K^* + K) - \tfrac{1}{4}(3\rho + \pi) = \Lambda - N$$
$$\text{(experimentally } 179 \approx 177) \tag{4e}$$

$$m_c - m_o = \tfrac{1}{4}(3D^* + D) - \tfrac{1}{4}(3\rho + \pi) = \Lambda_c - N$$
$$\text{(experimentally } 1356 \approx 1318) \tag{4f}$$

The presence of the factor $\frac{1}{2}$ in the expression for the spin-spin interaction of the quarks in the baryon can be checked without use of the ratio of the hadron volumes if we turn to the electromagnetic mass differences (see a forthcoming article [7]). Making the substitution for mesons $-g^2 \to -g^2 + e_1 e_2$ and for baryons $-g^2/2 \to -g^2/2 + e_i e_j$, we have

$$\frac{D_0^* - D_0}{D_+^* - D_+} = \frac{\Sigma_+^* - \Sigma_+}{\Sigma_-^* - \Sigma_-} = \frac{m_d}{m_u} + \frac{2}{3}\frac{e^2}{g^2}$$

and experimentally,

$$1 + (1.79 \pm 1.45) \times 10^{-2} \approx 1 + (2.03 \pm 0.67)^{-2} \tag{4g}$$

Taking for the parameters of Eqs. (1) and (2) the values

$m_o = 285 \qquad m_s = 463 \qquad m_c = 1621$

$\delta_M = 40 \qquad \delta_B = 230 \qquad b = 615$

we find the hadron masses, which are given below together with their experimental values. For the mesons,

	π	ρ	ω	K	K*	φ	D	D*	ψ
Formula	149	764	764	504	882.5	1024	1865	1973	3287
Experiment	138	773	783	494	892	1020	1865	2005	3105

and for the baryons,

	N	Δ	Σ	Λ	Σ*	Ξ	Ξ*	Ω	Σ_c	Λ_c
Formula	931	1239	1188	1109	1377	1335	1524	1677	2436	2267
Experiment	939	1232	1193	1116	1385	1318	1533	1672	2425	2257

In Ref. 2 we used formulas with a larger number of parameters than in Eqs. (1) and (2), namely,

$$M = a + \Sigma (m_s - m_o)|s| + \Sigma (m_c - m_0)|c| + b \Sigma \xi_i \xi_j \sigma_i \sigma_j \tag{5}$$

with different parameters for the mesons and baryons. The mass differences $D^* - D = 132$ MeV and $\Sigma_c - \Lambda_c = 161.5$ MeV predicted in Ref. 2 have been confirmed experimentally: respectively 140 and 168 MeV (see Refs. 5 and 6). However, the absolute mass values turned out to be somewhat higher than those predicted.

The best description of the masses of the D, D^*, and Λ_c (but not of the ψ) requires some change of the constants in Ref. 2, namely, for the mesons,

$$a = 614 \qquad b = 635 \qquad m_s - m_o = 179$$
$$m_c - m_o = 1356 \qquad \xi_s = 0.626 \qquad \xi_c = 0.22$$

and for the baryons,

$$a = 1085.5 \qquad b = 195.5 \qquad m_s - m_0 = 177$$
$$m_c - m_o = 1318 \qquad \xi_s = 0.605 \qquad \xi_c = 0.178$$

[ξ_c for the baryons is determined as $m_0/m_c = 285/(285 + 1318)$.]

With these constants the expected masses of the $s\bar{c}$ and $c\bar{s}$ mesons are 2083 for $J = 0$ and 2171 for $J = 1$. The $\psi - \eta''$ mass difference (for the vector and pseudoscalar particles of composition $c\bar{c}$), without allowance for the mixing with η and η', is 31 MeV.

The masses of the baryons turn out to be equal to the following values:

	$ooc(\Lambda_c \Sigma_c)$	$osc(\Xi_c)$	$ssc(\Omega_c)$	$occ(N_{cc})$	$scc(\Omega_{cc})$	$ccc(\Omega_{ccc})$
$J = 1/2$	2257/2417.5	2491.5/2582.5	2754	3688.5	3879	–
$J = 3/2$	2469.5	2624	2786	3740.5	3910.5	5044

The experimental value of the Σ_c mass is $\Sigma_c = 2425$ MeV. If we define

$$\xi_c = 1 - \frac{3}{2} \frac{\Sigma_c - \Lambda_c}{\Delta - N} = 0.14$$

then the calculated $\Xi_c'' - \Xi_c'$ mass difference increases somewhat:

$$\Xi_c'' - \Xi_c' = \frac{b}{2^{1/2}} [(\xi_s - \xi_c)^2 + (\xi_s - \xi_s \xi_c)^2 + (\xi_c - \xi_c \xi_s)^2]^{1/2}$$

$$= \begin{cases} 91 \text{ MeV} & \text{for } \xi_c = 0.178 \\ 96.7 \text{ MeV} & \text{for } \xi_c = 0.14 \end{cases}$$

The author expresses his gratitude to the participants in the seminar of the Theoretical Division at the Lebedev Institute for discussion of this question.

Note Added by the Author (April 1980)

In 1966 Zel'dovich and I proposed a linear mass formula uniformly describing mesons and baryons. That was the time of the SU_3 enthusiasm in which the u, d, and s quarks were used as a basis. Presented in the above-mentioned paper was the hypothesis of the dynamic nature of the mass difference $\Sigma - \Lambda$ due to the spin-spin interaction for oo quarks ("ordinary") and os quarks ("ordinary" with "strange" quarks). At the time, this was a rather original conjecture, as it ran against the "symmetry" philosophy of the day. In the papers of 1975 and the present paper (1979) these ideas are developed further. In this way the elementary semiqualitative dynamical explanation of hadron mass spectra is completed. This approach cannot replace more thorough calculations but is a good supplement to them. I would like to draw particular attention to the formula (3) which describes the mass splitting of Ξ'_c, Ξ''_c, Ξ^*_c baryons (of composition o, s, c) and its very simple derivation.

REFERENCES

1. Ya. B. Zel'dovich and A. D. Sakharov, *Yad. Fiz.* 4:395 (1966); *Sov. J. Nuclear Phys.* 4:283 (1967), trans. [Paper 19 in this volume].
2. A. D. Sakharov, *ZhETF Pis'ma* 21:554 (1975); *JETP Lett.* 21:258 (1975), trans. [Paper 20 in this volume].
3. J. Sapirstein, *The Color Magnetic Moment of a Quark*, SLAC-PUB-2352, June 1979.
4. Review of Particle Properties, in *Rev. Mod. Phys.* 48:No. 2, Part II (1976).
5. G. Goldhaber et al., *Phys. Rev. Lett.* 37:255 (1976).
6. C. Baltay et al., *Phys. Rev. Lett.* 42:1721 (1979).
7. A. D. Sakharov, *ZhETF* 79:350 (1980); *Sov. Phys. JETP* 52:175 (1980), trans. [Paper 22 in this volume].

Paper 22

An Estimate of the Coupling Constant Between Quarks and the Gluon Field

A simple estimate of the magnitude of the effective coupling constant between quarks and the gluon field is proposed, based on a consideration of the electric-charge dependence of the hadron mass differences for hadrons of identical composition. A comparison of the mass differences $\rho_+ - \pi_+$ and $\rho_0 - \pi_0$ yields $g^2/4\pi = 0.255^{+0.08}_{-0.05}$. A consideration of the mass differences for the D mesons and Σ baryons leads to a mutually consistent value $g^2/4\pi = 0.368^{+0.16}_{-0.08}$, which is larger than the value obtained from the $\rho - \pi$ system. The agreement between the results for the D and Σ may be considered as a confirmation of the assumption made in these estimates about the rank of the color group.

In this note we estimate the effective coupling constant of the quark-gluon interaction in hadrons (more precisely, its ratio to the coupling constant of the electromagnetic interaction) by considering the dependence of the mass difference between hadrons of identical composition on their electric charge. This method gives only a very rough estimate, but in our opinion is of some interest on account of its simplicity. A comparison of the results obtained for mesons and baryons makes it possible to verify the conjectured rank of the color group ($r = 2$ for the SU^c_3 group). We start from a linear mass formula for mesons and baryons proposed previously by Zel'dovich and the author [1], generalizing it to take into account the electromagnetic effects. The mass splitting between hadrons of identical composition,[†] according to Refs. 1 and

Source: Постоянная взаимодействия кварков с глюонным полем ЖЭТФ 79:350-353 (1980). Reprinted from *Sov. Phys. JETP* 52:175–176 (1980), translated by M. E. Mayer, with permission from American Institute of Physics.

[†]The previously calculated (Refs. 1 and 2) mass differences between hadrons of identical composition $\Sigma^* - \Sigma$, $\Xi^* - \Xi$, $\Sigma - \Lambda$, $D^* - D$, $\Sigma_c - \Lambda_c$ are in satisfactory agreement with experiment.

2, is described by the spin-spin interaction $H_{\sigma\sigma}$ of the quarks, which we interpret as the interaction between the gluonic quasi-magnetic moments $g/2m$ of the quarks, where g is the effective quark-gluon coupling constant and m is the quark mass.

This interpretation is confirmed by the fact that the empirical coefficients ξ introduced in Refs. 1 and 2, describing the weakening of the interaction of the s and c quarks with an adequate degree of accuracy (10%), are inversely proportional to the quark masses. Without taking account of the electromagnetic effects, we have for the mesons

$$H_{\sigma\sigma} = \frac{Cg^2}{V_M m_1 m_2} \sigma_1 \sigma_2 \tag{1}$$

where $\sigma_1 \cdot \sigma_2$ is the scalar product of the quark spins, V_M is the effective volume of the meson, $V_M^{-1} \sim |\psi(0)|^2$, and C is a constant. For baryons we have a similar expression of three terms, in which it is necessary to take into account an extra factor of $\frac{1}{2}$ stemming from the properties of the color group. The color charge for the putative SU_3^c color group of rank $r = 2$ is a two-dimensional vector. The scalar product of the charge vectors of a quark and antiquark making up a meson equals $-g^2$; in a baryon the charge vectors of different color quarks are arranged under angles of $120°$ in the charge plane and their scalar product equals $-\frac{1}{2}g^2$.

For a baryon we have (V_B is the effective volume of the baryon)

$$H_{\sigma\sigma} = \frac{Cg^2}{2V_B}\left(\frac{\sigma_1 \sigma_2}{m_1 m_2} + \frac{\sigma_2 \sigma_3}{m_2 m_3} + \frac{\sigma_3 \sigma_1}{m_3 m_1}\right) \tag{2}$$

The factor $\frac{1}{2}$ in Eq. (2) corresponds to the color group SU_3^c. In the more general case of the group SU_n^c we would have the factor $1/r = 1/(n-1)$ [the ratio of the radii of the inscribed and circumscribed spheres for the hypertriangle (simplex) in $(n-1)$-dimensional space].

In our previous notations [1,2], (m_0 is the mass of the nonstrange quark),

$$\frac{Cg^2}{V_M m_0^3} = b_M = \rho - \pi = 635 \text{ MeV}$$

$$\frac{Cg^2}{2V_B m_0^2} = b_B = \tfrac{2}{3}(\Delta - N) = 195.3 \text{ MeV}$$

We find that $V_B/V_M = \frac{3}{2}$, corresponding to the numbers of quarks in the baryon and meson. Such a ratio of effective volumes is quite plausible.

We generalize Eqs. (1) and (2) by adding to the interaction of the gluon moments the interaction of the Dirac magnetic moments, which are proportional to the quark charges. For the mesons we make the substitution

$$-g^2 \to -g^2 + e_1 e_2$$

and for the baryons we set

$$\frac{-g^2}{2} \to \frac{-g^2}{2} + e_i e_j$$

We will have for mesons,

$$H_{\sigma\sigma} = \frac{C\sigma_1\sigma_2}{V_M m_1 m_2}(g^2 - e_1 e_2) \tag{3}$$

and for baryons,

$$H_{\sigma\sigma} = \frac{C}{V_B}\left[\frac{\sigma_1\sigma_2}{m_1 m_2}\left(\frac{g^2}{2} - e_1 e_2\right) + \frac{\sigma_2\sigma_3}{m_2 m_3}\left(\frac{g^2}{2} - e_2 e_3\right) + \frac{\sigma_3\sigma_1}{m_3 m_1}\left(\frac{g^2}{2} - e_3 e_1\right)\right] \tag{4}$$

The mass differences between hadrons of identical composition will be determined according to these formulas by a change in the quantity $\sigma_i \cdot \sigma_j$; thus, for $D_+^* - D_+$ it is determined by a change of the quantity $\sigma_{\bar{d}} \cdot \sigma_c$ from the value $+\frac{1}{4}$ to the value $-\frac{3}{4}$, and the mass difference $\Sigma_+^* - \Sigma_+$ is determined by the change of $\sigma_u \cdot \sigma_s$ from the value $+\frac{1}{4}$ to the value $-\frac{1}{2}$. Taking into account the difference between the masses of the d and u quarks $m_d/m_u = 1 + \delta$, and introducing the notation $\varepsilon = e^2/g^2$, we get, up to quadratic terms,

$$\frac{\rho_0 - \pi_0}{\phi_+ - \pi_+} = \frac{1}{2}\left[\frac{1}{m_d^2}\left(g^2 + \frac{e^2}{9}\right) + \frac{1}{m_u^2}\left(g^2 + \frac{4e^2}{9}\right)\right]$$

$$\times \left[\frac{1}{m_d m_u}\left(g^2 - \frac{2e^2}{9}\right)\right]^{-1} = 1 + \frac{\varepsilon}{2} \tag{5.1}$$

(the numerator is the mean over the pairs $u\bar{u}$, $d\bar{d}$).

Similarly, we have obtained for other particles

$$\frac{K_0^* - K_0}{K_+^* - K_+} = 1 + \frac{\varepsilon}{3} - \delta \tag{5.2}$$

$$\frac{D_0^* - D_0}{D_+^* - D_+} = 1 + \frac{2\varepsilon}{3} + \delta \tag{5.3}$$

$$\frac{\Xi_0^* - \Xi_0}{\Xi_-^* - \Xi_-} = 1 + \frac{2\varepsilon}{3} + \delta \tag{5.4}$$

$$\frac{\Sigma_+^* - \Sigma_+}{\Sigma_-^* - \Sigma_-} = 1 + \frac{2\varepsilon}{3} + \delta \tag{5.5}$$

In arriving at the expressions (5.1) to (5.5) it was assumed that the Coulomb interaction of the quarks depends only on the composition of the hadrons, and that a change of the gluon interaction with the change of the spin function $\sigma_1 \cdot \sigma_2$ does not depend on the electric charge.

Making use of the experimental data $\rho_0 - \rho_+ = 4.5 \pm 2.3$ (here and in the sequel all masses are in MeV), $\pi_+ - \pi_0 = 4.60, \rho - \pi = 635$, we obtain

$$\frac{\varepsilon}{2} = (1.43 \pm 0.36) \times 10^{-2} \tag{6.1}$$

We have further

$$K_0^* - K_+^* = 4.1 \pm 0.6 \qquad K_0 - K_+ = 3.99 \pm 0.13 \qquad K^* - K = 398$$

$$\frac{\varepsilon}{3} - \delta = (0.025 \pm 0.15) \times 10^2 \tag{6.2}$$

$$D_+^* - D_0^* = 2.6 \pm 1.8 \qquad D_- - D_0 = 5.1 \pm 0.8 \qquad D^* - D = 140$$

$$\frac{2\varepsilon}{3} + \delta = (1.79 \pm 1.45) \times 10^{-2} \tag{6.3}$$

$$\Xi_-^* - \Xi_0^* = 3.3 \pm 0.7 \qquad \Xi_- - \Xi_0 = 6.4 \pm 0.6 \qquad \Xi^* - \Xi = 215$$

$$\frac{2\varepsilon}{3} + \delta = (1.45 \pm 0.32) \times 10^{-2} \tag{6.4}$$

$$\Sigma_-^* - \Sigma_+^* = 4.1 \pm 1.3 \qquad \Sigma_- - \Sigma_+ = 7.98 \qquad \Sigma^* - \Sigma = 192$$

$$\frac{2\varepsilon}{3} + \delta = (2.03 \pm 0.67) \times 10^{-2} \tag{6.5}$$

Making use of the data for ρ and π we obtain from (6.1), $g^2/4\pi = 0.255^{+0.081}_{-0.049}$.

Making use of the experimental data for K, D, and Σ, we obtain

$$\frac{g^2}{4\pi} = 0.368^{+0.16}_{-0.08} \qquad \delta = (0.63 \pm 0.43) \times 10^{-2}$$

Thus, we have obtained a simple (albeit insufficiently accurate) estimate of the effective coupling constant of the quark-gluon interaction. The distinction between the results obtained for the $\pi - \rho$ system and those for D and Σ is related to the dependence of the effective coupling constant on the size of the hadrons, which are different for the π and D. The large disagreement of the results for Ξ from all the other results cannot be understood, but it is still within the error limits. The approximate agreement of the results for the D meson ($\frac{2}{3}\varepsilon + \delta = 1.79 \times 10^{-2}$) and for the baryon ($\frac{2}{3}\varepsilon + \delta = 2.03 \times 10^{-2}$) confirms the existence of the factor $\frac{1}{2}$ in front of g^2 in Eq. (5.5), i.e., that the rank of the color group $r = 2$.

For a different rank ($r \neq 2, r = n - 1$, for the group SU_n^c) we obtain

$$\Delta = \frac{\Sigma_+^* - \Sigma_-}{\Sigma_-^* - \Sigma_-} - \frac{D_0^* - D_0}{D_+^* - D_+} = \frac{\varepsilon}{3}(r - 2)$$

This yields $\Delta \simeq 0.64 \times 10^{-2}$ for the group SU_4; experimentally $\Delta \simeq (0.24 \pm 1.6) \times 10^{-2}$, i.e., so far the accuracy is insufficient to exclude this possibility.

I would like to express my gratitude to the participants of the Seminar of the theoretical division of FIAN (the Lebedev Institute) for a discussion.

Note Added by the Author (April 1980)

The proposed method is based on consideration of the electromagnetic differences in masses and is semiquantitative in character. We believe that it is of some interest for its simplicity. The paper partly overlaps in content with the 1979 paper "Mass formula for mesons and baryons" [Paper 21 in this volume].

REFERENCES

1. Ya. B. Zel'dovich and A. D. Sakharov, *Yad. Fiz.* 4:395 (1966); *Sov. J. Nuclear Phys.* 4:283 (1967), trans. [Paper 19 in this volume].
2. A. D. Sakharov, *ZhETF Pis'ma* 21:554 (1975); *JETP Lett.* 21:258 (1975), trans. [Paper 20 in this volume].

Paper 23

Generation of the Hard Component of Cosmic Rays

The generation of the hard component of cosmic rays (mesons) by photons and nucleons is investigated in the relativistic region and near the generation threshold under two definite assumptions about the interaction of the nucleons.

I. INTRODUCTION

In this paper, the generation of the hard component of cosmic rays (mesons) is investigated in the framework of two possible assumptions about the interaction of nucleons which do not lead when applied to nuclear forces to a pole of the type $1/r^3$. The two assumptions are: (1) a scalar meson field, and (2) a pseudoscalar meson field possessing a "charge" (as opposed to "dipole") interaction with nucleons. These two forms of interaction are postulated in the theory of nuclear forces proposed by Tamm [1].

In Tamm's theory, nucleons at short distances interact in accordance with a $1/r$ law, which makes it possible to find the stationary states of a nucleus in the framework of the usual quantum-mechanical formalism. Accordingly, the calculation of cross sections in the present work does not lead to an anomalous growth of the cross section with energy. Moreover, at high energies the ratios of the cross sections of the nuclear processes to the cross sections of the corresponding electromagnetic processes decrease with energy. This "weakening of the nuclear interaction at high nucleon velocities" can be given a classical interpretation. The cross section found for the generation of mesons by protons is small and cannot be reconciled with the generally adopted opinion that the hard component of the cosmic rays is generated by primary protons.

Source: Генерация жесткой компоненты космических лучей, ЖЭТФ 17:686-697 **(1947).** Translated by J. B. Barbour.

Besides this relativistic problem, we also investigate the case when the energy of the primary particles is low, and we are at the "generation threshold." Here, one can expect large interference effects. However, in the case of generation of pseudoscalar mesons by photons, nuclear interference does not occur, and the cross section per nucleus for all angles and energies is proportional to Z and $A - Z$ for positive and negative mesons, respectively.

It is shown in the present paper that the threshold of meson generation accompanying scattering of nucleons by nuclei is in practice not μc^2 but $2\mu c^2$. This circumstance may, in particular, play a part in interpreting the spectrum of neutrons in the cosmic rays. This conclusion does not depend on the specific assumptions about the nuclear interaction. For the same process of meson generation accompanying the scattering of nonrelativistic nucleons, we find (without using the Born approximation) the ratio of the number of generated charged mesons to the number of generated neutral mesons. There are many more of the latter.

We use the following notation. The velocity of light c and Planck's constant $\hbar$ are taken equal to unity. Thus, we regard the velocity as a dimensionless quantity, and we measure the masses, energy, and momentum in units of reciprocal length. Further, M is the rest energy of the nucleon (i.e., proton or neutron); μ is the rest energy of the charged meson; μ_0 is the rest energy of the neutral meson; p_0, p_i, and p_1 are the momenta of a nucleon in the initial, intermediate, and final states; E_0, E_i, and E_1 are the energies of the nucleon in the same states; $\mathbf{k}_i$ and $\mathbf{k}_1$ are the momenta of a meson in the intermediate and final states; ω_i and ω_1 are the energies of the meson in the same states; ε is the momentum (and energy) of a photon; and ϑ is the proton-scattering angle. All the quantities so far introduced correspond to the coordinate system in which the center of mass is at rest. Further, E_L and ε_L are the energies of the primary particle in the coordinate system in which the other colliding particle is at rest before the collision. The electric charge and meson charge are measured in Heaviside units. Thus, $e^2 = 4\pi/137 = 1/10.8$.

We assume the existence of two kinds of meson field: an uncharged scalar ϕ_0 and a charged pseudoscalar ϕ. These have the following operators for their interaction with nucleons:

$$H'_0 = g\rho_3 \phi_0 \qquad V_0 = \sqrt{\frac{g^2}{2\omega_0}}\, u_2^* \rho_3 u_1$$

$$H = f\rho_2(P\phi + P^*\phi^*) \qquad V = \sqrt{\frac{f^2}{2\omega}}\, u_2^* \rho_2 P u_1 \qquad (1)$$

Here, g and f are dimensionless coupling constants whose values must be found experimentally, ρ_2 and ρ_3 are the usual Dirac matrices, and P is an operator which acts on the isotopic spin. If we denote by u_p and u_N the proton and neutron wave functions, respectively, then (by definition)

$$Pu_N = 0 \qquad P^*u_N = u_p \qquad (2)$$
$$Pu_p = u_N \qquad P^*u_p = 0$$

In what follows, we shall for simplicity not write out the operator P. In the second column of (1), we have written down the matrix elements for the emission of one meson.

In this paper, the cross sections are calculated in the Born approximation. We recall the basic formula for the differential cross section:

$$dS = \frac{2\pi\, d\rho}{|v_0'| + |v_0''|} VV^* \qquad (3)$$

For a simple and general derivation of this formula, see Heitler [2]. Here, $|v_0'|$ and $|v_0''|$ are the absolute magnitudes of the velocities of the colliding particles, and $d\rho$ is a function of the density and equal to the number of final states in the energy interval that satisfy the condition of periodicity in a unit cube and the law of conservation of momentum and charge.

If we introduce δ functions for the total momentum $\mathbf{P}$ and total energy E and denote the momenta of the particles in the final state by $\mathbf{p}_n$, then

$$d\rho = \frac{d\mathbf{p}_1}{(2\pi)^3} \cdots \frac{d\mathbf{p}_n}{(2\pi)^3} (2\pi)^3 \delta(\Delta E)\delta(\Delta \mathbf{P}) \qquad (4)$$

Eliminating the δ functions, we find expressions for different special cases. Throughout, we are interested in the center-of-mass system. We write down the expressions using a notation corresponding to their subsequent use.

1. In the case of two particles in the final state,

$$d\rho = \frac{k^2\, d\Omega}{(2\pi)^3(|v_1'| + |v_1''|)} \qquad (4a)$$

where k is the momentum of each of the particles, v_1' and v_1'' are the velocities in the final state, and $d\Omega$ is the element of solid angle.

2. In the case of three particles in the final state,

$$d\rho = \frac{k_1^2\, dk_1\, d\Omega_1}{(2\pi)^6} dp_{1y}'\, dp_{1z}' \frac{1}{|v_{1x}' - v_{1x}''|} \qquad (4b)$$

where k_1 and $d\Omega_1$ are the modulus of the momentum and the element of solid angle for the first particle (meson); $dp'_{1y}\, dp'_{1z}$ is the projection of the element of momentum space for the second particle (proton) onto the yz plane; $|v'_{1x} - v''_{1x}|$ is the projection of the difference between the velocities of the third and the second particle (two protons) onto the x axis.

By V in (3), we understand the matrix element of the first nonvanishing approximation calculated perturbatively. For example, when the first approximation of perturbation theory in which the interaction does not vanish is the third, we restrict ourselves to this approximation and write

$$V = \sum \frac{V_{A1} V_{12} V_{2F}}{\Delta_1 \Delta_2} \tag{5}$$

Here Δ_1 and Δ_2 are the energy defects in the intermediate states:

$$\Delta_1 = \Sigma E_{1i} - \Sigma E_0 = \Sigma E_{1i} - \Sigma E_1$$
$$\Delta_2 = \Sigma E_{2i} - \Sigma E_0 = \Sigma E_{2i} - \Sigma E_1$$

The summation in (5) is over all possible chains of intermediate states.

The matrix elements V_{A1}, V_{12}, and V_{2F} for the individual virtual quantum transitions are nonvanishing if the total momentum and charge are conserved. In calculating these matrix elements, it is important to take into account the antisymmetry of the nucleon wave functions with respect to the set of arguments: the momentum **p**, the spin σ, and the isotopic spin τ.

II. GENERATION OF THE HARD COMPONENT OF COSMIC RAYS BY PHOTONS

There are three different possible generation mechanisms in which the primary particle is a photon:

$$h\nu + P \to \mu_+ + N \tag{6a}$$

$$h\nu + N \to \mu_- + P \tag{6b}$$

$$h\nu + P \to \mu_0 + P \tag{6c}$$

We shall investigate in detail the process (6a). Besides the interaction (1), this process involves the interaction (7) of the proton with light and the interaction (8) of the charged meson field ϕ with light:

$$H' = -\int e(A\alpha)\, d\mathbf{x} \tag{7}$$

FIELD THEORY AND ELEMENTARY PARTICLES

$$H' = - ie \int d\mathbf{x}\, A \left(\frac{\partial \phi^*}{\partial \mathbf{x}} \phi - \frac{\partial \phi}{\partial \mathbf{x}} \phi^* \right) + e^2 \int A^2 \phi \phi^*\, d\mathbf{x} \qquad (8)$$

The quadratic term in (8) is omitted in later calculations.

We must take into account three chains of intermediate states (Fig. 1a shows the momenta of all the states that are important for the process and Fig. 1b shows the scheme of the quantum transitions). Each arrow corresponds to a quantum transition, and the circle indicates the interaction responsible for it.

Let a_ε, a_k, and b_k be the operators of creation of a photon and positive and negative mesons, and $\mathbf{e}_\lambda$ be the unit polarization vector of the photon:

$$A = \sum_{\lambda,\varepsilon} \sqrt{\frac{1}{2\varepsilon}}\, (a_\varepsilon e^{-i\varepsilon \mathbf{k}} + a_\varepsilon^* e^{i\varepsilon \mathbf{x}})$$

$$\phi = \sum_k \sqrt{\frac{1}{2\omega}}\, (a_k e^{-i\mathbf{kx}} + b_k^* e^{i\mathbf{kx}}) \qquad (9)$$

$$\phi^* = \sum_k \sqrt{\frac{1}{2\omega}}\, (a_k^* e^{i\mathbf{kx}} + b_k e^{-i\mathbf{kx}})$$

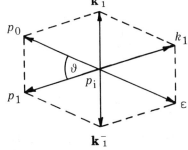

a

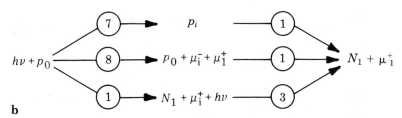

b

Figure 1

Using (1), (5), (7), and (9), we can find the total matrix element. We have

$$V = V_\mathrm{I} + V_\mathrm{II} + V_\mathrm{III} \tag{10}$$

$$V_\mathrm{I} = -\frac{ef}{\sqrt{4\omega_1\varepsilon}} \sum \frac{(u_1^*\rho_2 u_i)(u_i^*\alpha_\lambda u_0)}{\Delta_i^\mathrm{I}}$$

where

$$\Delta_i^\mathrm{I} = \pm M - (E_0 + \varepsilon) \tag{11}$$

$$V_\mathrm{II} = -ie[(-i\mathbf{k}_i^- + i\mathbf{k}_1^+)\mathbf{e}_\lambda]\sqrt{\frac{1}{16\omega_1\varepsilon\omega_i^2}}\,\frac{f(u_1^*\rho_2 u_0)}{\Delta_i^\mathrm{II}}$$

where

$$\Delta_i^\mathrm{II} = \omega_1 + \omega_i - \varepsilon$$

$$V_\mathrm{III} = -ie[(+i\mathbf{k}_i^+ + i\mathbf{k}_1^+)\mathbf{e}_\lambda]\sqrt{\frac{1}{16\omega_1\varepsilon\omega_i^2}}\,\frac{f(u_1^*\rho_2 u_0)}{\Delta_i^\mathrm{III}} \tag{12}$$

where

$$\Delta_i^\mathrm{III} = \varepsilon + \omega_i - \omega_1$$

From the conservation laws,

$$\varepsilon + \sqrt{M^2 + \varepsilon^2} = \omega_1 + \sqrt{\omega_1^2 - \mu^2 + M^2}$$

from which it follows that $\omega_1 = \varepsilon + \tfrac{1}{2}\mu^2/(E_0 + \varepsilon)$. In what follows, we shall ignore the second term and assume

$$\omega_1 = \varepsilon \qquad E_0 = E_1$$

$$\Delta_i^\mathrm{II} = \Delta_i^\mathrm{III} = \omega_i = \sqrt{\mu^2 + p_0^2 + p_1^2 - 2p_0 p_1 \cos\vartheta}$$

To transform (10), we use the well-known device (see Heitler [2]) of multiplying the numerator and denominator by $M + (\varepsilon + E_0)$. We obtain [remembering that $M^2 - (\varepsilon + E_0)^2 = -2\varepsilon(E_0 + \varepsilon)$]

$$V = \frac{ef[u_1^*\rho_2(\beta M + \varepsilon + E_0)\alpha_\lambda u_0]}{4\varepsilon^2(E_0 + \varepsilon)} + \frac{ef(u_1^*\rho_2 u_0)(\mathbf{k}_1\mathbf{e}_\lambda)}{\omega_i\varepsilon} \tag{13}$$

To calculate the cross section, we must sum VV^* over the neutron spin σ_1 and average over the initial spin σ_0 of the proton and the polarization λ (again see Ref. 2). Using (3) and (4a) (and setting $p_0 = \varepsilon$ and $p_1 = k_1 = \sqrt{\varepsilon^2 - \mu^2}$), we obtain

$$dS = d\Omega \frac{e^2 f^2 p_1}{16(2\pi)^2 p_0 (E_0 + \varepsilon)^2}$$

$$\times \left[\frac{E_0 + \varepsilon \cos \vartheta}{E_0 + \varepsilon} + \frac{2p_1^2(p_0^2 + p_1^2 - 2p_1 p_0 \cos \vartheta) \sin^2 \vartheta}{(p_0^2 + p_1^2 - 2p_1 p_0 \cos \vartheta + \mu^2)^2} \right.$$

$$\left. + \frac{2p_1^2 \sin^2 \vartheta}{p_0^2 + p_1^2 - 2p_0 p_1 \cos \vartheta + \mu^2} \right] \tag{14}$$

In limiting cases, the formula simplifies. For $k_1 \gg \mu$,

$$dS = d\Omega \frac{e^2 f^2}{16(2\pi)^2 (E_0 + \varepsilon)^3} \left(\frac{E_0 + \varepsilon \cos \vartheta}{E_0 + \varepsilon} + 2 + 2 \cos \vartheta \right) \tag{15}$$

Integration over the angles gives

$$S = \frac{e^2 f^2 (3E_0 + 2\varepsilon)}{16\pi (E_0 + \varepsilon)^3} \tag{16}$$

For $k_1 \ll \mu$,

$$dS = d\Omega \frac{e^2 f^2}{16(2\pi)^2 M^2} \sqrt{1 - \frac{\mu^2}{\varepsilon^2}} \tag{17}$$

$$S = \frac{e^2 f^2}{16\pi M^2} \left(1 - \frac{\mu^2}{\varepsilon^2}\right)^{3/2} \tag{18}$$

Taking $f^2/4\pi = 0.5$ for an estimate, we obtain $S_{\max} = 1.5 \times 10^{-29}$ cm^2.

We recall that (16) corresponds to the center-of-mass system. We go over to the laboratory system. The total cross section does not change under the Lorentz transformation, and the transformation of the energy is given by the formula

$$(\Sigma E_L)^2 - (\Sigma p_L)^2 = \text{inv} = (\Sigma E_0)^2 \tag{19}$$

After transformations, we obtain in our case

$$E_0 = (M + \varepsilon_L) \left(\frac{M}{M + 2\varepsilon_L}\right)^{1/2} \qquad \varepsilon = \varepsilon_L \left(\frac{M}{M + 2\varepsilon_L}\right)^{1/2}$$

As was pointed out by Feinberg [3], in a number of cases it is important to take into account interference effects in the nucleus. In nonstationary processes, the particles in a nucleus act independently (and give a total cross section for one nucleus proportional to the number of particles in the nucleus) only when the momentum s transferred to the nucleus satisfies the inequality

$$sa \gg 1 \tag{20}$$

Here, a is the "mean distance" of the particles in the nucleus from each other, $a = 1/\mu$ in order of magnitude, and s in our notation is $s = |\varepsilon - \mathbf{k}|$.

However, in the case of generation of a pseudoscalar meson by a proton, we must also expect the absence of appreciable interference effects in the cases when the condition (20), or even the weaker condition

$$sR \gg 1 \tag{20a}$$

where R is the radius of the nucleus, is not satisfied. To prove this, we consider the matrix element (13) in the nonrelativistic approximation. It can be transformed into

$$V = U_1^*[\mathbf{A}\mathbf{e}_\lambda + B(p_0 + p_1)\boldsymbol{\sigma}]U_0 \tag{21}$$

Here, U are two-row wave functions of Pauli's theory, $\mathbf{A}$ and B are coefficients which depend on the angles and the energies, and $\boldsymbol{\sigma}$ is the spin vector (two-row Pauli operator); the expression (21) has different signs depending on the spin orientation. Therefore, the interference terms in the expression for the cross section must vanish if protons of both spin orientations are present in the nucleus in equal numbers. The cross section per nucleus is found to be proportional to Z (for positive mesons) and to the number of neutrons in the nucleus for negative mesons.

We give without derivation the cross section for the process (6c) in the nonrelativistic approximation (Θ is the angle between $\mathbf{k}_1$ and $\mathbf{e}_\lambda$):

$$dS_c = \frac{e^2 g^2}{4(2\pi)^2 M^2}\left(1 - \frac{\mu^3}{\varepsilon^2}\right)^{1/2} \cos^2 \Theta \tag{22}$$

$$S_c = \frac{e^2 g^2}{12\pi M^2}\left(1 - \frac{\mu^2}{\varepsilon^2}\right)^{3/2} \tag{23}$$

Setting $g^2/4\pi = 0.25$, we obtain $S_{\max} = 3 \times 10^{-30}$ cm^2.

The expression (22) can be interpreted classically. Consider a proton which executes oscillations due to the field of the incident light wave. The spherical meson wave that propagates from it in all directions can be found by means of retarded potentials and is (in the wave zone)

$$\phi \sim \frac{\cos \Theta}{r} \sin(\varepsilon t - r\sqrt{\varepsilon^2 - \mu^2}) \tag{24}$$

From $\varepsilon < \mu$, the meson wave does not penetrate into the wave zone;

FIELD THEORY AND ELEMENTARY PARTICLES 247

more precisely, only waves with multiple frequencies $n\varepsilon$ and very low intensity penetrate to the wave zone.[†] In the case of generation of charged mesons we do not find such a simple classical interpretation.

III. GENERATION OF THE HARD COMPONENT OF COSMIC RAYS WHEN HEAVY PARTICLES ARE DECELERATED: THE RELATIVISTIC CASE

We show that within the framework of the assumption made about the nuclear interaction there is a common cause of rapid decrease of all cross sections with the energy. We consider the square of the matrix elements of (1), averaging over the spins of the initial and final states (see Heitler [2]):

$$\frac{1}{2} \underset{\sigma_0 \sigma_1}{SS} |(u_1^* \rho u_0)|^2 = \frac{1}{8 E_0 E_1} \operatorname{Tr} \rho(H_0 + E_0)\rho(H_1 + E_1)$$

$$= \frac{1}{2 E_0 E_1} (E_0 E_1 \pm M^2 - p_0 p_1 \cos \vartheta)$$

Here, the plus sign corresponds to the case of the scalar interaction $\rho = \rho_3$, and the minus sign to the case of the pseudoscalar interaction $\rho = \rho_2$.

We have the identity

$$2(E_0 E_1 - M^2 - p_0 p_1 \cos \vartheta) = -(E_0 - E_1)^2 + p_0^2 + p_1^2 - 2 p_0 p_1 \cos \vartheta$$

Finally, we have

$$\frac{1}{2} SS |(u_1^* \rho_2 u_0)|^2 = \frac{1}{4 E_0 E_1} [(\Delta p)^2 - (\Delta E)^2] \quad (25a)$$

$$\frac{1}{2} SS |(u_1^* \rho_3 u_0)|^2 = \frac{1}{4 E_0 E_1} [(\Delta p)^2 - (\Delta E)^2] + \frac{M^2}{E_0 E_1} \quad (25b)$$

We see that the matrix elements of (1) corresponding to small momenta of the virtual meson decrease rapidly with increasing energy E.

This circumstance plays a decisive part in the behavior of fast particles of fields of the type considered. Note that the "weakening of interactions" affects not only the cross sections for collision processes

[†]In the language of quantum transitions, these waves correspond to "multiple" processes with the absorption of several primary photons.

but also the occurrence of stationary levels. As an example, we consider the stationary scalar field $g^2\rho_3/r$. The field is analogous to an electric field of the form Ze^2/r. In an electric field, it is well known that because of relativistic effects a solution for the S state does not exist for $Ze^2 > 1$. In a scalar field of the type considered, such a solution exists for any g^2, as was pointed out to me by I. E. Tamm in discussion. Namely, the lowest level is determined by the formula

$$E = \frac{M}{\sqrt{1 + g^4}} \qquad (26)$$

(instead of $E = M\sqrt{1 - e^4 Z^2}$ in the electric case).

In the nonrelativistic approximation, the properties of the two fields are identical. In particular, (26) goes over into the expression in Schrödinger theory:

$$E = M - \frac{g^4 M}{2}$$

In the case of the scalar interaction, the weakening effect admits a classical interpretation. Namely, the matrix ρ_3 in classical theory corresponds to the quantity $\sqrt{1 - v^2}$, which tends to zero as $v \to 1$; the scalar field $g\phi_0 = U(x,t)$ can be described by the Lagrange function

$$\mathcal{L} = -\sqrt{1 - v^2}(U + M)$$

and the Hamilton function

$$H = \sqrt{p^2 + (U + M)^2}$$

In particular, the expression (26) can be derived by means of the Bohr–Sommerfeld quantization rules if one sets $U = g^2/r$.

In the special case of *Bremsstrahlung* processes accompanying the exchange interaction of heavy particles, this difference between the scalar field and the electric field was first noted by Wang [4]. However, Wang's arguments are based entirely on the von Weizsäcker–Williams method, and we have therefore felt it desirable to discuss the question from other points of view.[†] We consider the typical process of genera-

---

[†] In Ref. 4, Wang finds that the equivalent spectrum of the scalar nuclear field of a relativistic proton contains $\xi = E_L/M$ times fewer scalar pseudoquanta of maximal energy than the corresponding electromagnetic spectrum found by von Weizsäcker. The profile of the spectrum is also different from the electromagnetic spectrum. Using the cross section for the scattering of a neutral scalar meson by a nucleon, we were able to derive the expression (35) by means of the equivalent spectrum found by Wang (up to a numerical factor).

Wang's results differ from ours, since the scattering cross section for the charged scalar meson which he adopts has a different energy dependence.

tion of scalar neutral mesons in the case of the scattering of protons by scalar nuclear forces. As can be seen by comparing (25a) and (25b), the process with the participation of the pseudoscalar meson cannot be more probable in order of magnitude.

We set ourselves the task of estimating an upper limit for the "energy loss cross section" S_E (see Heitler [2]). We make the following simplifications in these calculations:

1. We take into account only states with positive energy as intermediate states. This will not involve a large error, since in states with negative energy the energy defect is very large (for the chain considered below, Δ_1 and Δ_2 are of the order of E_0).

2. We shall take into account only one chain of intermediate states (instead of 24), namely (Fig. 2),

$$p'_0 + p''_0 \to p'_i + \mu_i + p''_0 \to p'_i + p''_1 \to p'_1 + p''_1 + \mu_1$$

Among the total number of 24, only 16 chains can make a contribution to the matrix element of the same order as this one (these are the chains for which a virtual meson with energy comparable to E is present only in one intermediate state). Of these 16, only four chains play a part for mesons emitted (approximately) in the direction of the primary proton. Only such mesons play a part in the calculation of S_E. If the four equal matrix elements were added, we would obtain an increase in the cross section by a factor of 16. Thus, taking into account only one chain, we make an error of not more than a factor of 16. The matrix element corresponding to our chain is

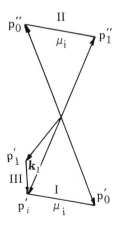

Figure 2

$$V = \sqrt{\frac{g^6}{8\omega_i^2 \omega_1}} \frac{V_1^2 V_2}{\Delta_1 \Delta_2}$$

$$V_1 = u_0'''^* \rho_3 u_1'' \qquad V_2 = u_i'^* \rho_3 u_1' \tag{27}$$

3. For V_1 and V_2, we use the approximate expressions corresponding to the first term in (25b). One can show that the second term in (25b) plays a part in a small region of angles and does not change the order of magnitude of the results for very high energies:

$$V = \frac{1}{2E} \sqrt{(\Delta \mathbf{p})^2 - (\Delta E)^2} \tag{28}$$

4. For the energy defects, we use approximate expressions. We denote the angle between the direction of the momentum of the meson and the proton in the final state by Ψ. We shall assume that $\Psi \ll 1$ and $\mu \ll k_1 \ll E$. We use the expansion

$$\sqrt{\mathbf{p}^2 + M^2} = p + \frac{M^2}{2p} \tag{29}$$

We have

$$\Delta_2 = E_i' - E_1' - \omega_1 = -\frac{k_1}{2}\left(\frac{M^2}{E^2} + \Psi^2\right) \tag{30}$$

We denote by δ the change in the energy of the second proton. Since $\delta = E_i' - E_0' = E_1'' - E_0''$ and $\delta = \Delta_2/2$, we have $\Delta_1 = \omega_i + \delta = \Delta_2/2 + \omega_i$.

It is expedient to estimate the cross section separately for two regions of angles. The first is the region where $\Delta_2/2 \approx \omega_i$. In this region, the angle ϕ between $\mathbf{k}_i$ and $\mathbf{p}_0$ is small. Therefore, as in the case of (30),

$$\Delta_1 = \omega_i + E_i' - E_0 = \frac{k_i}{2}\left(\frac{M^2}{E^2} + \phi^2\right)$$

The second region is where $|\Delta_2/2\Delta| \ll \omega_i$, and we can assume

$$\Delta_1 = \omega_i \tag{31}$$

We shall not make calculations for the first region. We note only that in this region it is impossible to use (28), since the second term in the expression (25b) plays the principal part in the cross section. In the integrated cross section, region II plays a more important part than region I (at high energies).

We calculate V_2 for the second region (the last transition in our chain). On the basis of (29), $\Delta E = k_1 \cos \Psi$ and $\Delta p = k_1$. On the basis of (28), $V_2 = \Psi k_1/2E$ (here $\sin \Psi$ is replaced by Ψ). We calculate V_1 for the second region (the second and the first transition in our chain). Here $\Delta E \ll \Delta p$, and $\Delta p \approx \omega_i$. We have $V = \omega_i/2E$. On the basis of (5), (27), (30), and (31), we arrive at the expression

$$\frac{1}{4} S_{\sigma_0} S_{\sigma_0'} S_{\sigma_1} S_{\sigma_1'} \frac{g^6}{2^7} \frac{\Psi^2}{E^6 \omega_1 \left(\frac{M^2}{E^2} + \Psi^2\right)^2} \tag{32}$$

We substitute (32) in (3), using (4b) and the approximate equations

$$v'_{1x} - v''_{1x} = 2 \qquad dp'_y \, dp'_z = E^2 \, d\Omega'$$

$$dS - \frac{g^6 \, d\Omega' \, dk_1 \, k_1 \, d\Omega_1 \, \Psi^2}{2^9 (2\pi)^5 E^4 \left(\frac{M^2}{E^2} + \Psi^2\right)^2} \tag{33}$$

There is no point in integrating exactly over the angles in (33), since the employed approximations are valid in the region of small angles:

$$\int \frac{\Psi^2 \, d\Omega_1}{(M^2/E^2 + \Psi^2)^2} \cong \int_0^Z \frac{\Psi^3 \, d\Psi}{(M^2/E^2 + \Psi^2)^2} \cong 2\pi \left(\ln \frac{ZE}{M} - \frac{1}{2}\right)$$

Here, Z is of order unity. The integration over Ω' gives the factor 4π.

The fact that we have found the scattering of protons to be isotropic shows the necessity of taking into account the interference between the chains and the symmetry of the wave functions. However, we shall not do this here, and for a rough estimate we take into account the other chains by means of the factor 16:

$$dS = \frac{g^6}{2(4\pi)^3} \frac{k_1 \, dk_1}{E^2} \frac{\ln(E/M)}{E^2} \tag{34}$$

Here, k_1 may be either positive or negative (we have mesons emitted "upward" and "downward"). However, only the mesons emitted in the direction of the primary proton are important for the energy loss. We go over to the laboratory coordinate system, and for the momentum of the generated meson introduce the notation l and for the momentum and energy of the primary proton, p_L and E_L. Applying (19), we obtain

$$E_0^2 = \tfrac{1}{2}(m^2 + ME_L) \qquad \frac{l}{k_1} = \frac{p_L}{p_0} \qquad \text{for } p_0 \gg M, k_1 \gg \mu \tag{35}$$

$$dS = \frac{1}{4} \left(\frac{g^2}{4\pi}\right)^3 \frac{l \, dt}{E_L^2} \frac{\ln(E_L/M)}{ME_L}$$

The energy-loss cross section is determined by (36). Here, N is the number of nucleons per unit volume, x is the coordinate in the direction in which the primary particles move, and ΔE_L is the energy loss in one collision:

$$-\frac{dE_L}{dx} = E_L N S_E \qquad S_E = \frac{1}{E_L} \int_0^{E_L} dS\, \Delta E_L \qquad (36)$$

In our case,

$$\Delta E_L \cong l$$

$$S_E = \frac{1}{12}\left(\frac{g^2}{4\pi}\right)^3 \frac{\ln(E_L/M)}{E_L M} \qquad (37)$$

Setting $g^2/4\pi = 1$ for an estimate, we obtain for $E_L = 10^{-10}$ eV the result $S_E = 8 \times 10^{-30}$ cm^2 (per nucleon of the nucleus) and $S_E = 2.5 \times 10^{-28}$ cm^2 (per nucleus of air).

If one takes the generally accepted view of the origin of the mesons in cosmic rays, according to which protons, which are decelerated through meson generation, form the primary component, then for such a small cross section there would be more primary protons on the surface of the earth than mesons. A cross section guaranteeing impenetrability of the atmosphere for the primary protons must exceed the one we have calculated by a factor of at least several hundred.

IV. GENERATION OF THE HARD COMPONENT OF COSMIC RAYS WHEN NUCLEONS WITH ENERGY NEAR THE GENERATION THRESHOLD ARE SCATTERED BY NUCLEI

We determine the meson generation threshold. Let M be the mass of the primary particles (nucleons) and M_0 the mass of a nucleus. Applying (19), we obtain for the limiting case when all the particles are at rest in the final state in the center-of-mass system,

$$(M_0 + E_L)^2 - p_L^2 = (M_0 + M + \mu)^2 \qquad p_L^2 = E_L^2 - M^2$$

$$T_0 = E_L - M = \mu\left(1 + \frac{M}{M_0}\right) + \frac{\mu^2}{2M_0} \qquad (38)$$

In the case of air, this quantity exceeds μ by 4%. However, it must be borne in mind that the coupling of the nucleons in the nucleus is so weak that the cross section has a certain value for the energies for which meson generation on the free nucleon is possible. In this case, $M_0 = M$ and $T_0 = E_L - M = 2\mu$.

Let us estimate the part played by coupling of the nucleons in the nucleus. In a nucleus, the nucleons are localized in a region of order $a = 1/\mu$. In the momentum space, the corresponding wave function $\Psi(\mathbf{p})$ decreases rapidly when $p > \mu$. We consider the collision of a primary nucleon with momentum $\mathbf{p}_0$ and kinetic energy T_0 with the nucleon of a nucleus. In the calculation of the matrix element of meson generation, only a certain part Γ of the space of momenta $\mathbf{p}$ of the nucleon plays a part; it is necessary to have

$$M + E_L \geq \sqrt{(2M + \mu)^2 + (\mathbf{p}_0 - \mathbf{p})^2}$$

with $\mathbf{p}$ within Γ. The influence of coupling on the cross section can be characterized in a rough estimate by the "kinematic factor" W:

$$W(p_0) = \int_{\Gamma(p_0)} |\Psi(\mathbf{p})|^2 \, d\mathbf{p}$$

The main contribution to this integral is made by the region of intersection of Γ with the region P where $|p| \leq \mu$. Suppose the regions $\Gamma(p_0)$ and P touch. The condition for this is

$$M + E_L = \sqrt{(2M + \mu)^2 + (p_0 - \mu)^2}$$

The solution of this equation gives $T_0 = E_0 - M = 1.55\mu$. At lower energies, P and Γ do not have common points. However, even at somewhat higher energies, up to 2μ, the kinematic factor W is small (less than 0.5).

It is possible that this result could play a part in the interpretation of the spectrum of slow neutrons in the cosmic rays.

In a recent paper, Powell [5], asserts that the majority of the cosmic ray neutrons that give rise to nuclear disintegration have energy less than 200 MeV. To explain this number, which is equal to twice the meson rest energy, Powell assumes: (1) that high-energy neutrons lose energy rapidly on meson generation and are therefore seldom observed as the sources of nuclear distintegrations, and (2) that mesons are generated in pairs.

It is clear that if $2\mu c^2$ is the threshold for generation of a single meson, it is possible to explain the facts without invoking the second, far reaching assumption.

There is little point in obtaining definite expressions for the cross section near the threshold in the Born approximation because of the large value of the coupling constants. However, we can estimate the ratio of the number of charged mesons to the number of neutral mesons. We consider the nucleon-nucleon system, taking into account only scalar (neutral) nuclear forces (in Tamm's theory, these forces play a decisive part in the nonrelativistic region). We can regard the generation of mesons in a collision of nucleons as the result of a transition to the continuous spectrum of this system. We shall ignore the difference between the coupling constants for the emission of scalar and pseudoscalar mesons, and also the difference that arises from taking into account the symmetry properties of the wave functions in the two cases. Then the required ratio is given by the ratio of the squares of the matrix elements (25a) and (25b). In the nonrelativistic approximation, we obtain (taking into account $\Delta p = k_1$)

$$\tfrac{1}{2} SSV_{ps}^2 = \frac{k_1^2}{4M^2} \tag{39a}$$

$$\tfrac{1}{2} SSV_s^2 = 1 \tag{39b}$$

Therefore, if k_1 is of order μ, we have several hundred times fewer charged mesons than neutral mesons.

I should like to express my gratitude to Professor I. E. Tamm, who suggested to me the subject of this paper and gave great assistance in its preparation.

REFERENCES

1. I. Tamm, *J. Phys. of the U.S.S.R.* 9:499 (1945).
2. W. Heitler, *The Quantum Theory of Radiation* [Russian translation of the first edition, Oxford (1936)], Moscow-Leningrad, GTTI (1940).
3. E. L. Feinberg, *J. Phys. of the U.S.S.R.* 5:176 (1941).
4. F. S. Wang, *Z. f. Phys.* 115:7 (1940).
5. W. M. Powell, *Phys. Rev.* 69:9, 385 (1946).

Paper 24

Interaction of the Electron and the Positron in Pair Production

The influence of the interaction of the components on the differential probability of pair production is studied.

I.

In calculations of the probability of pair production, allowance is never made for the interaction of the components of the pairs. Some authors (Heitler [1]) are of the opinion that it is impossible to solve this problem in the framework of hole theory.

Our method is based on the following remarks:

1. The interaction of the components has a significant influence on the differential probability only when the components have a small relative velocity in the final state. Therefore, such interaction can be treated in the center-of-mass system of the electron and positron as a simple Coulomb interaction e^2/r.

2. This interaction cannot be repeated as a perturbation but must be taken into account in the calculation of the eigenfunctions of the electron-positron system, as can be seen from the correction factor (8) which we obtain for the probability, the electron charge occurring in this expression in an essentially transcendental manner.

II.

We use below a system of units in which $\hbar = c = m$ (mass of the electron) $= 1$.

Source: Взаимодействие электрона и позитрона в процессе образования пар, ЖЭТФ 18:631-635 (1948). Translated by J. B. Barbour.

The remaining notation is as follows: $e = 137^{-1/2}$ is the electron charge, E_+, E_- and $\mathbf{p}_+, \mathbf{p}_-$ are the energy and momenta of the positron and electron in the laboratory coordinate system, $\mathbf{k}_+$ and $\mathbf{k}_-$ are the momenta of the positron and electron in the coordinate system in which $\mathbf{k}_+ \approx \mathbf{k}_-$ (center-of-mass system), and v is the relative velocity of the electron and the positron. It is readily seen that

$$v = \sqrt{1 - (E_+ E_- - \mathbf{p}_+ \mathbf{p}_-)^{-2}} \approx \sqrt{p^2 \vartheta^2 + (|\mathbf{p}_+| - |\mathbf{p}_-|)^2 E^{-2}} \qquad (1)$$

Here, ϑ is the angle between the electron and the positron; the approximate expression for v corresponds to the case $v \ll p/E$ (i.e., $\mathbf{p}_+ \to \mathbf{p}_-$).

The wave function describing the electron and the positron is $\Psi_{\mathbf{p}_+' \mathbf{p}_-'}(q)$, where q denotes the arguments of the wave function; in the usual formulation of hole theory, this is the set of coordinates of the electrons of all occupied levels.

We use a system of eigenfunctions that go over into plane waves at the infinity of the coordinate space of the electron and positron. The indices $\mathbf{p}_+'$ and $\mathbf{p}_-'$ are the momenta corresponding to these plane waves (the momenta at infinity); $\Psi_0(q)$ is the wave function that describes the vacuum; V is the transition matrix element, and $\widetilde{V}$ is the matrix element calculated without allowance for the interaction of the components. Quite generally, we use the tilde to denote the values of quantities calculated without allowance for the interaction of the components, in contrast to the "exact" values.

The required correction factor for the differential probability $d\omega$ is

$$T = \frac{d\omega}{\widetilde{d\omega}} = \left|\frac{V}{\widetilde{V}}\right|^2 \qquad (2)$$

III.

In perturbation theory calculations, the Hamiltonian is represented as a sum of two terms: $H = H_0 + H_1$. The first term is used to calculate the eigenfunctions $H_0 \Psi = E_0 \Psi$, and the second is used to calculate the matrix element. For first-order processes,

$$V_{\mathbf{p}_+' \mathbf{p}_-'} = \int \Psi_0^* H_1 \Psi_{\mathbf{p}_+' \mathbf{p}_-'}^* \, dq \qquad (3)$$

(The following section is devoted to processes of higher order.) We shall assume that the interaction W of the components is included in H:

$$H_0 = \widetilde{H}_0 + W$$

The eigenfunctions Ψ of the operator H are linear combinations of the eigenfunctions $\widetilde{\Psi}$ of the operator $\widetilde{H}_0$. We have

$$\Psi_{\mathbf{p}'_+\mathbf{p}'_-} = \int d^3\mathbf{p}'_+ \, d^3\mathbf{p}'_-(\mathbf{p}_+\mathbf{p}_-|c|\mathbf{p}'_+\mathbf{p}'_-)\widetilde{\Psi}_{\mathbf{p}_+\mathbf{p}_-} \qquad (4)$$

(for brevity, the spin variables are omitted here and below). In Eq. (4), c is some unitary singular matrix very close to a δ matrix. Indeed, in the limiting case $e \to 0$,

$$c \to \widetilde{c} = \delta(\mathbf{p}_+ - \mathbf{p}'_+)\delta(\mathbf{p}_- - \mathbf{p}'_-)$$

The exact form of c will be found later.

Substituting (4) and (3) and reversing the order of integration over q and p, we obtain

$$V_{\mathbf{p}'_+\mathbf{p}'_-} = \int d^3\mathbf{p}_+ \, d^3\mathbf{p}_-(\mathbf{p}_+\mathbf{p}_-|c|\mathbf{p}'_+\mathbf{p}'_-)\widetilde{V}_{\mathbf{p}_+\mathbf{p}_-} \qquad (5)$$

Because of the δ-like nature of c, we can take $\widetilde{V}$ in front of the integral sign and write

$$J = \frac{V}{\widetilde{V}} = \int d^3\mathbf{p}_+ \, d^3\mathbf{p}_- \, c \qquad (6)$$

Further, we can go over to the center-of-mass system of the electron and the positron:

$$\mathbf{p}_+ \to \mathbf{k}_+ \qquad \mathbf{p}'_+ \to \mathbf{k}'_+, \ldots \qquad \mathbf{k}'_+ \approx -\mathbf{k}'_-$$

$$(\mathbf{k}_+\mathbf{k}_-|c_k|\mathbf{k}'_+\mathbf{k}'_-) = \sqrt{\lambda\lambda'}(\mathbf{p}_+\mathbf{p}_-|c|\mathbf{p}'_+\mathbf{p}'_-)$$

Here, $\lambda(\mathbf{p}_+, \mathbf{p}_-)$ and $\lambda'(\mathbf{p}'_+, \mathbf{p}'_-)$ are the Jacobians of the transformation, and the factor $\sqrt{\lambda\lambda'}$ ensures that the unitarity of the matrix c is conserved:

$$\int d^6\mathbf{p} \, c^*c = \delta(\mathbf{p}'_+ - \mathbf{p}''_+)\delta(\mathbf{p}'_- - \mathbf{p}''_-)$$

$$\int d^6\mathbf{k}(\mathbf{k}|c_k^*|\mathbf{k}')(\mathbf{k}|c|\mathbf{k}'') = \delta(\mathbf{k}'_+ - \mathbf{k}''_+)\delta(\mathbf{k}'_- - \mathbf{k}''_-)$$

In calculating J [Eq. (6)], we can assume that $\lambda' \approx \lambda$ (because of the δ-like nature of c) and write

$$J = \int \lambda^{-1} c_k \, d^6\mathbf{p} = \int c_k \, d^6\mathbf{k}$$

Finally, we can make the Fourier transformation

$$\int c_k \, d^6\mathbf{k} = (2\pi)^3 (00|c_x|\mathbf{k}'_+\mathbf{k}'_-) \qquad (7)$$

Here, c_x is the wave function in the coordinate space of the electron and positron, which satisfies (in the nonrelativistic approximation) the Schrödinger equation

$$-\tfrac{1}{2}(\Delta_+ + \Delta_-)c_x - \left(\frac{e^2}{r}\right) c_x = \tfrac{1}{2}(\mathbf{k}'^2_+ + \mathbf{k}'^2_-)c_x$$

We introduce relative coordinates in the usual manner and solve the resulting equation with reduced mass $\frac{1}{2}$ by separating the variables in parabolic coordinates (cf. Bethe's solution to the problem of electron scattering [2]). We obtain (normalization to volume $d^3\mathbf{k}'_+ \times d^3\mathbf{k}'_-$)

$$c_x = \frac{F(i\varepsilon, 1, l[(\mathbf{k}'_+ - \mathbf{k}'_-)(\mathbf{x}_+ - \mathbf{x}_-) + |\mathbf{k}'_+ - \mathbf{k}'_-| |\mathbf{x}_+ - \mathbf{x}_-|])}{|F(i\varepsilon,1,i\infty)|} \tilde{c}_x$$

$$\tilde{c}_x = (2\pi)^{-3} \exp i(\mathbf{k}'_+ \mathbf{x}_+ + \mathbf{k}'_- \mathbf{x}_-)$$

Here, F is the hypergeometric function, $\varepsilon = e^2/v$, where v is the relative velocity of the electron and the positron in the final state (1), and $\mathbf{x}_+$ and $\mathbf{x}_-$ are radius vectors.

On the basis of (7),

$$J = |F(i\varepsilon, 1, i\infty)|^{-1} = \left(\frac{2\pi\varepsilon}{1 - e^{-2\pi\varepsilon}}\right)^{1/2}$$

Finally, on the basis of (2),

$$T = J^2 = \frac{2\pi\varepsilon}{1 - e^{-2\varepsilon\pi}} \qquad (8)$$

IV.

The generalization of this derivation to processes of second (and higher) order is not difficult. The matrix element is calculated in the form of a sum (or integral, for generality) over so-called intermediate states $\mathbf{p}^i$. Instead of (3) for first-order processes, we obtain for second-order processes

$$V_{\mathbf{p}'} = \int d^3\mathbf{p}^i_+ d^3\mathbf{p}^i_- (\int \Psi_0^* H_1 \Psi_i \, dq)(\int \Psi_i^* H_1 \Psi_{\mathbf{p}'} \, dq) \Delta_i^{-1} \qquad (9)$$

where Δ_i is the change in the energy in the intermediate state compared with the initial state. The final function $\Psi_{\mathbf{p}'} = \Psi \mathbf{p}'_+ \mathbf{p}'_- (q)$ occurs in this formula linearly. The interaction of the components of the pair in the final state is taken into account in exactly the same way as in the case of first-order processes, and again leads to (8).

We now consider the influence of the interaction of the components in the intermediate state on the differential probability. Even without calculation it is clear that the point at which the relative velocity v^i in the intermediate state vanishes cannot be a singular point for the correction factor T (in contrast to the point of vanishing of v in the final state, which is such a singular point). The point is that the relative

FIELD THEORY AND ELEMENTARY PARTICLES

velocity v^i in the intermediate state is not a relativistically invariant quantity. If $v^i = 0$ in one frame of reference, then it will be nonzero in other frames of reference.

We shall show by a typical example that the interaction in the intermediate state is in fact unimportant. Let us consider pair production by a photon in the field of a nucleus, i.e., the term of the matrix element V due to the chain

$$\mathbf{p}_\gamma \to \mathbf{p}_+ + \mathbf{p}_-^i \to \mathbf{p}_+ + \mathbf{p}_- + \mathbf{q}$$

where $\mathbf{p}_\gamma$ is the momentum of the photon, and $\mathbf{q}$ is the momentum transferred to the nucleus. Here $v^i = 0$ for $\mathbf{p}_+ = \mathbf{p}_\gamma/2$, and it is obvious that this condition is relativistically noninvariant.

Substituting (4) in (9), we obtain

$$V = \int d\mathbf{p}_+^i \, d\mathbf{p}_-^i \, \Delta_i^{-1} [\int \widetilde{V}_1 \, (\mathbf{p}^1|c|\mathbf{p}^i \, \delta(\mathbf{p}_+^1 + \mathbf{p}_-^1 - \mathbf{p}_\gamma) d\mathbf{p}^1]$$
$$\times [\int \widetilde{V}_2(\mathbf{p}^2|c^*|\mathbf{p}^i) \, \delta \, (\mathbf{p}_+^2 - \mathbf{p}_+) \, d\mathbf{p}^2]$$

Taking the slowly varying factors in front of the integral sign, we obtain $V = \widetilde{V}J$, where $\widetilde{V} = \widetilde{V}_1\widetilde{V}_2/\widetilde{\Delta}_i$, and

$$J = \int d\mathbf{p}^i \, d\mathbf{p}^1 \, d\mathbf{p}^2 \, \delta \, (\mathbf{p}_+^2 - \mathbf{p}_+)\delta(\mathbf{p}_+^1 + \mathbf{p}_-^1 - \mathbf{p}_\gamma)(\mathbf{p}^1|c|\mathbf{p}^i)(\mathbf{p}^2|c^*|\mathbf{p}^i) \quad (10)$$

Because of momentum conservation, the matrix c contains a δ-like factor, which it is expedient to split off. We set $\mathbf{p}_+^i + \mathbf{p}_-^i = \mathbf{p}_\sigma^i$, $\mathbf{p}_+^1 + \mathbf{p}_-^1 = \mathbf{p}_\sigma^1$, etc. We have

$$(\mathbf{p}_+^1\mathbf{p}_-^1|c|\mathbf{p}_+^i\mathbf{p}_-^i) = \delta(\mathbf{p}_\sigma^1 - \mathbf{p}_\sigma^i)(\mathbf{p}_-^1|d|\mathbf{p}_-^i) \quad (11)$$

where the new matrix d is also unitary:

$$\int d\mathbf{p}_-^i(\mathbf{p}_-^1|d|\mathbf{p}_-^i)(\mathbf{p}_-^2|d^*|\mathbf{p}_-^i) = \delta(\mathbf{p}_-^1 - \mathbf{p}_-^2) \quad (12)$$

Substituting (11) in (10), we find

$$J = \int d\mathbf{p}_-^i \, d\mathbf{p}_-^1(\mathbf{p}_-^1|d|\mathbf{p}_-^i)(\widetilde{\mathbf{p}}_-^i|d^*|\mathbf{p}^i)$$

which is equal to unity by virtue of (12). Thus, in the approximation in which the entire theory is constructed ($\widetilde{V}$ and Δ taken in front of the integral sign), the interaction in the intermediate state is indeed unimportant, which agrees with the invariance requirements.

V.

Hitherto we have ignored spin and relativistic effects. Do they influence our results? Equation (5) remains exactly the same, but the form of the matrix c is changed somewhat, and in the summation over the spin variables it is necessary to take into account the dependence of $\widetilde{V}$ on the spins. However, this last circumstance is unimportant, since the spin is conserved in the Coulomb interaction of slow particles, so that $\widetilde{V}$ can be taken in front of the sign of the summation over the spins.

Equations (6) and (7) do not hold, since $(00|c_x|)$ becomes infinite. Instead of taking $\widetilde{V}$ in front of the integral in (5), we can on the basis of a well-known theorem make a Fourier transformation of $\widetilde{V}$ and c:

$$\int d^6\mathbf{p}\widetilde{V}_p c_p = \int d^6\mathbf{x} V_x c_x \tag{13}$$

Here V_x is some δ-like function smeared over the region of space responsible for the pair production. (In the case of pair production as a result of a nuclear transition with forbidden emission of photons, the initial angular momentum of the nucleus is $J = 0$; V_x corresponds to the oscillations of the Coulomb potential, and $V_x \neq 0$ within the nucleus. See Sakharov [3], and also Oppenheimer [4], and Yukawa and Sakata [5].)

The function c_x in (13) has a "weak" pole with degree of order 137^{-2} (by analogy with the function of a single electron in a Coulomb field). Since $\widetilde{V}$ is smeared over a region of order of the radius R of the nucleus or more, and c_x differs from its nonrelativistic value in regions of order of the electron radius $r_0 \ll R$, we can in the calculation of (13) use instead of the exact values of c_x its nonrelativistic value at the coordinate origin. We again arrive at (7).

VI.

We note that the region of quantitative applicability of Eq. (8) is limited to medium Z (the charge of the nucleus) and relativistic velocities of the electrons and positrons on account of the "Born" treatment of the Coulomb field of the nucleus. In contrast, the relative velocity of the components may be arbitrarily small, since to treat the interaction we have not used the Born approximation (in contrast to Ref. 6, in which Rudnitskii studied annihilation; of course, our results also apply to annihilation).

This paper forms part of my dissertation. I am pleased to express my gratitude to my supervisor Professor I. E. Tamm.

REFERENCES

1. W. Heitler, *The Quantum Theory of Radiation* [Russian translation of the first edition, Oxford (1936)], Moscow-Leningrad, GTTI (1940).
2. H. A. Bethe, *Quantum Mechanics of the Simplest Systems* [Russian translation], Moscow-Leningrad (1935). [Translation of Bethe's 1933 *Handbuch*

der Physik (Vol. 24) article "Quantenmechanik der Ein- und Zwei-Elektronenprobleme."]
3. A. D. Sakharov, *Dissertation*, P. N. Lebedev Physics Institute (1947).
4. J. R. Oppenheimer, *Phys. Rev. 59*:216(A) (1941).
5. Yukawa and Sakata, *Proc. Phys.-Math. Soc. Jpn. 17*:10 (1935).
6. Rudnitskii, *ZhETF 7*:1303 (1937).

Commentary: Stephen L. Adler[*]

Induced Gravitation

As recently reviewed by Weinberg [1], in the conventional grand unified picture gravitation appears on a quite different footing from the weak, strong, and electromagnetic interactions of the matter fields. The total dynamics, in the usual formulation, is governed by an action functional[†]

$$S = \int d^4x \sqrt{-g}\, (\mathscr{L}_{\text{matter}} + \text{counterterms} + \mathscr{L}_{\text{gravitation}}) \tag{1}$$

with $\mathscr{L}_{\text{matter}}$ a renormalizable Lagrangian density, containing only dimensionless coupling constants, and $\mathscr{L}_{\text{gravitation}}$ the Einstein–Hilbert gravitational Lagrangian:

$$\mathscr{L}_{\text{gravitation}} = \frac{1}{16\pi G} R \tag{2}$$

with R the curvature scalar. Since the coupling constant $(16\pi G)^{-1}$ appearing in the gravitational action has the dimensionality $(\text{mass})^2$, quantization of the gravitational part of Eq. (1) leads, as is well known, to a nonrenormalizable field theory. Furthermore, in the conventional view there is no mechanism for tying the gravitational mass scale set by $G^{-1/2}$ to the unification mass scale of the matter interactions. Gravitation thus appears as a phenomenon quite outside the usual framework of theoretical ideas on which elementary particle theory is based.

I will describe in this lecture some recent work which offers the possibility of both relating the gravitational and unification mass scales, and of permitting the formulation of a quantum theory of gravitation as a renormalizable field theory. The basic premise of what

---
[*] The Institute for Advanced Study, Princeton, New Jersey
[†] In this lecture I use the metric and curvature conventions of C. W. Misner, K. S. Thorne, and J. A. Wheeler, *Gravitation*, W. H. Freeman, San Francisco (1973).

follows, first suggested by A. D. Sakharov [2], is that the Einstein action is not a fundamental action at all, but rather is an induced effect—in Sakharov's words the "metrical elasticity of space"—resulting from quantum fluctuations of the matter fields. Sakharov's argument starts from the renormalized matter action[†]

$$\tilde{S} = \int d^4x \sqrt{-g}\, \tilde{\mathcal{L}}$$
$$\tilde{\mathcal{L}} = \mathcal{L}_{\text{matter}} + \text{counterterms} \tag{3a}$$

from which one obtains the vacuum action functional

$$<\tilde{S}>_0 = \int d^4x \sqrt{-g}\, <\tilde{\mathcal{L}}>_0 \tag{3b}$$

Because quantum effects lead to nonlocal interactions with the space-time curvature, the expectation $<\tilde{\mathcal{L}}>_0$ cannot be related to its flat space-time value by the equivalence principle. Instead, we have a formal expansion in even powers of derivatives of the metric,

$$<\tilde{\mathcal{L}}>_0 = <\tilde{\mathcal{L}}>_0^{\text{flat space-time}} + \frac{1}{16\pi G_{\text{ind}}} R + O(R^2, R_{\alpha\beta}R^{\alpha\beta}) \tag{4}$$

with the matter quantum fluctuations inducing an effective Einstein–Hilbert action. The coefficient $\propto G_{\text{ind}}^{-1}$ appearing in this action is unfortunately infinite; in Sakharov's formulation, it is given by the quadratically divergent integral

$$\frac{1}{16\pi G_{\text{ind}}} \propto \int k\, dk = \text{quadratic divergent} \tag{5}$$

Clearly, to carry this very interesting suggestion further, conditions must be found which guarantee that G_{ind}^{-1} is finite, and is unambiguously calculable. The first point to be made is that from the modern viewpoint of dimensional or ζ-function regularization (I restrict myself throughout to regularizations[‡] involving continuation in a *dimensionless* parameter), quadratic divergences vanish, and so the leading divergence structure of Eq. (5) has the logarithmically divergent form

$$\frac{1}{16\pi G_{\text{ind}}} = a\,\kappa^2 \times \log \text{divergence} + b\kappa^2 \tag{6}$$

where log divergence is a power series in $(n-4)^{-1}$ in dimensional

---

[†] I use a tilde (-) to indicate renormalized quantities.
[‡] I am implicitly assuming that all reasonable regularizations involving continuation in a dimensionless parameter give the same result for G_{ind}.

regularization, with κ a characteristic mass scale of the matter theory. When a is nonzero, the finite residue b is ill defined, since it can be changed by redefinition of the logarithmically divergent term. If, however, a vanishes, then this ambiguity in b disappears, and the induced gravitational constant becomes both finite and well defined.

I will show, in what follows, that the condition for a to vanish is that the matter theory exhibit the phenomenon of spontaneous, dynamical scale symmetry breaking. A connection between spontaneous scale-breaking and gravitation was suggested recently by several authors [3] in the context of a Higgs-type model:

$$S = \int d^4x \sqrt{-g}\, [\tfrac{1}{2}\varepsilon\phi^2 R - V(\phi^2) \\ + \text{kinetic terms, counterterms, other fields}] \tag{7}$$

If quantized around the symmetric vacuum $\phi = 0$, Eq. (7) gives no order R term in the action. Suppose, however, that V has a stable minimum away from $\phi = 0$,

$$V'(\kappa^2) = 0$$

$$V''(\kappa^2) > 0 \tag{8}$$

so that spontaneous symmetry breakdown occurs, with the true vacuum being either $\phi = \kappa$ or $\phi = -\kappa$. The symmetry breakdown clearly induces an effective gravitational action of the form

$$S_{\text{gravitation}} = \int d^4x \sqrt{-g}\, \tfrac{1}{2}\varepsilon\kappa^2 R \tag{9}$$

The coefficient ε in Eq. (9) is still a free parameter of the model, divergent (but renormalizable) when quantum corrections are included.

To get the possibility of a finite induced gravitational action, it is necessary to exclude scalar fields (as well as bare, or kinetic, mass terms) from the theory, as is done in the so-called "heavy color" or "technicolor" unified models [4]. In these models all particle masses are calculable in terms of a unification mass κ, and the gist of what I have to say is that $G_{\text{ind}}^{-1/2}$ behaves then as simply another mass and is also calculable in terms of κ. That is, *in renormalizable field theory, if all particle masses are calculable, Newton's constant* G_{ind} *is also calculable* [5].

A formal proof of this assertion uses the Bogoliubov–Parasiuk–Hepp–Zimmermann (BPHZ) renormalization algorithm, which states that operators which can mix under renormalization are polynomials of the same canonical dimension, and of the same symmetry type (both Lorentz and internal), formed from all the bare masses and bare fields present in the Lagrangian and their derivatives. It is well known that this algorithm implies that particle masses will be calculable only if the following two conditions are satisifed:

1. All bare masses are zero (so that scale-symmetry breaking is spontaneous, rather than explicit).
2. No scalar fields are present; the theory must contain only spin-$\frac{1}{2}$ fermion and spin-1 gauge fields (hence scale-symmetry breaking must occur from a dynamical mechanism, not from an explicit Higgs sector).

Let us now examine the behavior of a theory satisfying conditions 1 and 2 when embedded in a curved background space-time manifold. Making the expansion of Eq. (4), we must ask whether $(16\pi G_{\text{ind}})^{-1} R$ can mix with any divergent quantities under renormalization; if it can, it will in general be a divergent quantity itself. Now according to the BPHZ algorithm stated above, the induced gravitation term mixes under renormalization with operators of the form $O_2 R$, with O_2 a canonical-dimension-2, Lorentz scalar, gauge-invariant operator. However, the restrictions 1 and 2 needed for calculability of particle masses imply that there are *no* such operators O_2; the only dimension-2 Lorentz scalars are of the form $b_\mu^A b^{A\mu}$, with b_μ^A a gauge potential, but this operator is not gauge invariant (and hence has a different internal symmetry structure from $G_{\text{ind}}^{-1} R$). Thus the theory contains no counterterms which mix with $(16\pi G_{\text{ind}})^{-1} R$, and hence G_{ind} is calculable [5], instead of being merely renormalizable.

The basic idea just outlined has been confirmed in calculations by Hasslacher and Mottola [6] and by Zee [6]. Hasslacher and Mottola calculate G_{ind} for an instanton gas in pure Yang–Mills gauge theory, and find an ultraviolet convergent, nonvanishing effect. Their expression contains terms of both signs, suggesting that the sign of G_{ind} in the ultimate grand unified theory will be determined by details of the scale-symmetry breaking mechanism. Zee has evaluated a Feynman graph model, in which Pauli–Villars regulators simulate the effect of dynamical scale breaking. He finds a nonzero effect, with a sign which is infrared dominated. (Earlier calculations of G_{ind} using regulators had been given by Akama et al. [7], and by Sakharov himself in two papers [8,9], with results very similar to those found by Zee.) The structure of the calculations of Ref. 6, in particular the appearance in both of the trace of the stress-energy tensor, suggests that it should be possible to do the curved space-time manipulations in a formal way, giving an explicit formula for the induced gravitational constant involving flat space-time quantities only.

This, in fact, is readily done, as follows [10]. In terms of the renormalized matter Lagrangian $\tilde{\mathcal{L}}$ of Eq. (3), the renormalized matter-stress-energy tensor is defined by[†]

$$\tilde{T}^{\mu\nu} = 2(-g)^{-1/2} \frac{\delta}{\delta g_{\mu\nu}} [(-g)^{1/2} \tilde{\mathcal{L}}] \tag{10}$$

Applying the metric variation appearing in Eq. (10) to the expansion

$$<\tilde{\mathcal{L}}>_0 = <\tilde{\mathcal{L}}>_0^{\text{flat space-time}} + \frac{1}{16\pi G_{\text{ind}}} R + \cdots \tag{11}$$

and contracting with $g_{\mu\nu}$ gives an equivalent expansion of the vacuum expectation of the trace of the stress-energy tensor,

$$<\tilde{T}^\mu_\mu>_0 = <\tilde{T}^\mu_\mu>_0^{\text{flat space-time}} + \frac{1}{8\pi G_{\text{ind}}} R + O(R^2) \tag{12}$$

Hence we can extract G_{ind} by calculating the change in $<\tilde{T}^\mu_\mu>_0$ produced by space-time curvature. To do this, let us use the familiar formula valid for a general Lagrangian variation:[‡]

$$\delta<\tilde{T}^\mu_\mu(0)>_0 = i \int d^4x \{<T(\delta[\sqrt{-g}\,\tilde{\mathcal{L}}(x)]\,\tilde{T}^\mu_\mu(0))>_0$$
$$- <\delta[\sqrt{-g}\,\tilde{\mathcal{L}}(x)]>_0 <\tilde{T}^\mu_\mu(0)>_0\} \tag{13}$$
$$= i \int d^4x <T(\delta[\sqrt{-g}\,\tilde{\mathcal{L}}(x)]\,\tilde{T}^\mu_\mu(0))>_{0,\text{connected}}$$

Since we wish to calculate a single number, it suffices to do the calculation in the special case of a conformally flat, constant curvature metric, which in a suitable inertial frame has the form

$$g_{\mu\nu}(x) = \eta_{\mu\nu}(1 - \tfrac{1}{24}Rx^2 + \cdots) \tag{14}$$

Hence, taking $\delta\tilde{\mathcal{L}}$ in Eq. (13) to be the Lagrangian variation induced by the metric change

$$\delta g_{\mu\nu}(x) = -\eta_{\mu\nu}\tfrac{1}{24}Rx^2 \tag{15}$$

and using Eq. (10) a second time gives

$$\frac{1}{8\pi G_{\text{ind}}} R = \delta <\tilde{T}^\mu_\mu(0)>_0$$

$$= \tfrac{1}{2} i \int d^4x \sqrt{-g}\,(-\tfrac{1}{24}Rx^2) <T(\tilde{T}^\lambda_\lambda(x)\,\tilde{T}^\mu_\mu(0))>_{0,\text{connected}}$$
$$+ O(R^2) \tag{16}$$

[†] I will work only to order R, omitting order R^2 counterterms which give rise to the curvature contribution to the conformal trace anomaly.

[‡] In order R^2, there will also be an intrinsic variation $\delta\tilde{T}^\mu_\mu(0)$ arising from the order R^2 counterterms.

We can now divide by R and take the flat space-time limit, giving a formula for Newton's constant,

$$\frac{1}{16\pi G_{\text{ind}}} = \frac{i}{96} \int d^4x (x^{0^2} - \mathbf{x}^2) <T(\tilde{T}_\lambda^\lambda(x)\tilde{T}_\mu^\mu(0))>_{\substack{\text{flat space-time}\\\text{0,connected}}} \quad (17)$$

Let us now examine the ultraviolet convergence properties of this formula. To eliminate the formal quadratic divergence which is always present, we must interpret Eq. (17) as the dimensional continuation limit[†]

$$\frac{1}{16\pi G_{\text{ind}}} = \frac{i}{96} \lim_{n \to 4} \int d^n x\,(-x^2)<T(\tilde{T}_\lambda^\lambda(x)\,\tilde{T}_\mu^\mu(0))>_{\substack{\text{flat space-time}\\\text{0,connected}}} \quad (18)$$

Poles at $n = 4$ can arise only from terms of the form

$$<T(\tilde{T}_\lambda^\lambda(x)\,\tilde{T}_\mu^\mu(0))>_{\substack{\text{flat space-time}\\\text{0,connected}}} = \cdots + <O_2>_{\substack{\text{flat space-time}\\\text{0,connected}}}$$
$$\times (x^2)^{-3} \times \text{logs} + \cdots \quad (19)$$

in the perturbative operator product expansion of the T product. As before O_2 is any gauge-invariant scalar operator of canonical dimension 2. Again the hypothesis of dynamical scale invariance breakdown implies that no such O_2's exist in the theory, and hence the dimensional continuation limit exits.

Note that although the classicial Lagrangian of Eq. (3) is conformally invariant, the quantum theory is not, and so low energy matrix elements of $\tilde{T}_\mu^\mu(x)$ will be nonvanishing. Hence Eq. (18) gives an ultraviolet-convergent, and in general nonvanishing, induced gravitational constant.

To summarize, Sakharov's idea that the Einstein–Hilbert action is an induced quantum effect may be realizable within a very interesting class of quantum field theories. Outstanding questions for the future research are (1) getting a definitive calculation of G_{ind} in some of the unified models currently being studied [the simplest version of this problem would be to get a calculation of G_{ind} in a pure $SU(n)$ gauge theory], and (2) examining the implications of an induced order-R action for the problem of quantizing gravitation—this problem will assume a totally different character if Sakharov's vision proves correct [11].

REFERENCES

1. S. Weinberg, *Rev. Mod. Phys.* 52:515 (1980).
2. A. D. Sakharov, *Dokl. Akad. Nauk. SSSR* 177:70 (1967); *Sov. Phys. Dokl.*

[†] Note that in Misner–Thorne–Wheeler conventions, $x^2 = \mathbf{x}^2 - x^{0^2}$.

12:1040 (1968), trans. [Paper 14 in this volume]. See also O. Klein, *Phys. Scr.* 9:69 (1974).
3. P. Minkowski, *Phys. Lett.* 71B:419 (1977). A. Zee, *Phys. Rev. Lett.* 42:417 (1979). L. Smolin, *Nucl. Phys. B160*:253 (1979). E. M. Chudnovskii, *Teor. i. Mat. Fiz.* 35:398 (1978); *Theoret. and Math. Phys.* 35:538 (1978), trans.; F. Englert, C. Truffin, and R. Gastmans, *Nucl. Phys. B117*:407 (1976).
4. L. Susskind, *Phys. Rev. D20*:2619 (1979). S. Weinberg, *Phys. Rev. D13*:974 (1976), *D19*:1277 (1979); or the $m_0 = 0$ version of S. L. Adler, *Phys. Lett. 86B*:203 (1979).
5. S. L. Adler, *Phys. Rev. Lett.* 44:1567 (1980).
6. B. Hasslacher and E. Mottola, *Phys. Lett.* 95B:237 (1980). A. Zee, *Phys. Rev. D23*:858 (1981).
7. K. Akama et al., *Progr. Theoret. Phys.* 60:868 (1978).
8. A. Sakharov, The vacuum quantum fluctuations in curved space and the theory of gravitation, in *Gravitation and Field Theory*, Institute of Applied Mathematics, Moscow (1967). Paper 15 in this volume, trans.
9. A. Sakharov, *Teor. i. Mat. Fiz.* 23:178 (1975); *Theoret. and Math. Phys.* 23:435 (1975), trans. [Paper 16 in this volume].
10. S. L. Adler, *Phys. Lett.* 95B:241 (1980).
11. For further discussion and references, see S. L. Adler, *Rev. Mod. Phys.* 54(3):729 (1982).

*Commentary: Harry J. Lipkin**

Elementary Particles

A few months ago one of my colleagues showed me a copy of a recent manuscript by Andrei Sakharov, written at his exile in Gor'kii and describing his recent research in theoretical high energy physics. The paper had been translated from Russian into broken English by a friend of Sakharov's in the U.S. who was not a physicist and who had no way to judge the scientific value of the work. I was asked to look it over and tell him whether it was of interest to the physics community and whether it should be published or distributed in some form. To my great surprise, I found the work very close to my own recent research. In fact, he had independently obtained many of the same results that I had published during the past two years. I have discussed this work in invited lectures at the annual Washington meeting of the American Physical Society last April and at international conferences in the U.S., Canada, and Europe during the summer. Since learning of Sakharov's new unpublished work, I always mention that my results have also been obtained independently by him. The reaction of the audience is generally one of amazement. Everyone asks, "How can he achieve this under his present conditions?"

Andrei Sakharov is known to be a brilliant and highly original physicist who has made interesting contributions to theoretical high energy physics a number of years ago. But the present work is not the kind done by a great genius in isolation. It is an updating of previous work making use of the recent theoretical and experimental developments which have taken place all over the world. Somehow Sakharov has managed to learn about them and keep up to date. Everyone knows of his recent political activities and his troubles and exile in Gor'kii.

*Physics Department, The Weizmann Institute of Science, Rehovot, Israel

They wonder how he has been able to remain informed and alert to those new results which have bearing on his previous ideas.

This work applies some of the latest ideas on the structure of fundamental particles to a particular approach developed by Sakharov in 1966. It happened that a group of my colleagues at the Weizmann Institute of Science in Rehovot, Israel developed almost the identical methods at the same time and obtained very similar results. But because of the Mideast tensions leading to the Six Day War in June 1967 and the breaking of diplomatic relations between Israel and the U.S.S.R., there was not very good communication between us at that time. I did not learn about Sakharov's work until many years later. During the past two years, Sakharov and I both independently noticed that recent developments and new knowledge now allowed a significant improvement in this old 1966 work, and obtained new results which again were almost identical.

It is remarkable that Andrei Sakharov has been able to produce such work under his present conditions.

I have been free to go to the best laboratories in the world and to discuss my work with the leading physicists in the outstanding centers of high energy research. In this way I learn immediately about new results. Sakharov is sitting in Gor'kii with no good library and no contact with the main centers of physics. In one of his manuscripts he refers to a book with the statement, "I do not remember the exact name of the book, and in Gor'kii I have nowhere to look." Yet he is aware of the latest results and knows how to fit them in with his previous work to draw new and interesting conclusions.

This recent work has now been translated into proper scientific English by experts in the U.S. and is available to the scientific community as an unauthorized translation. It is a disgrace that Sakharov is unable to communicate his contributions to the international community of science by the normal channels of participation in international conferences and free publication in scientific journals.

I first present a summary of his work in nontechnical language, and later quote the technical details.

Matter is made of molecules. Molecules are made of atoms. Atoms are made of nuclei and electrons. Nuclei are made of neutrons and protons. This view of the structure of matter is supported by many experiments in which matter is broken up into its constitutent neutrons, protons and electrons. These are then isolated, accelerated into beams, and shot at one another, thus enabling detailed investigations to be made of the forces between them and how they behave. Looking inside the proton and neutron is much more difficult. Physicists now believe that the protons and neutrons are built of even smaller building blocks, called quarks. But nobody has succeeded in breaking up the proton or neutron and seeing isolated quarks.

Physicists have attempted to knock quarks out of a proton in the same way as they break up nuclei into neutrons and protons by shooting tremendous amounts of energy at the proton. But this energy did not break the proton up into quarks. The energy was simply transformed into new matter and made more protons and other particles, including new kinds of particles not known before. In analyzing these experiments, physicists have concluded that the forces that bind quarks together into protons must be very strong. It must take much more energy to break up a proton than it does to create many protons. It may be a long time before we can get enough energy into a proton to break it up into isolated quarks. Some physicists even believe that this can never happen.

But if we cannot break the proton up into its quarks, how do we know that it really is made of quarks and how can we study the forces between quarks? This question was investigated by Andrei Sakharov and Ya. B. Zel'dovich in 1966, and independently by P. Federman, H. R. Rubinstein, and I. Talmi in Israel. The basic idea was very simple. Quarks came in three different types, now called "flavors." Quarks of two flavors, called "up" and "down" quarks, are found in the proton and neutron. There are also heavier quarks, called "strange" quarks, found in other particles. All the different kinds of particles created when protons are bombarded with energy are supposed to be built from these same quarks in their three flavors rearranged in different ways. Nothing was then known about the forces between quarks. Sakharov and Zel'dovich simply assumed that these forces must be the same no matter how the quarks were arranged in different particles. In this way they were able to write down a formula which predicted the masses of particles containing different numbers of up, down, and strange quarks arranged in different ways. The experimental values of the masses agreed with their formula, thus giving support to the theory that the particles were made of quarks. But there were uncertain factors in the formula because the nature of the forces between quarks was completely unknown.

There was also another difficulty which perplexed many physicists. If quarks behave like ordinary particles, a general principle called "Fermi statistics" states that two quarks with exactly the same flavor cannot occupy the same point in space at the same time. But the standard model for the proton had two quarks of the same flavor at the same point. This bothered Sakharov and Zel'dovich, and they concluded (incorrectly as we now know) that the proton could not be made only of three quarks and probably consisted of five quarks. The discussion of this five-quark model for the proton is the main point of their 1966 paper. However, there is also a section on the three-quark model which is still valid today. This was not noticed as the paper was disregarded because of its emphasis on the five-quark model. The Weizmann Institute group ignored the statistics problem completely, assuming (correctly it turned out) that this would be solved eventually in a different way. In fact the solution accepted today had already been proposed by O. W. Greenberg. In Greenberg's model, quarks have another property in addition to flavor, now called "color." The three quarks in the proton were required to have different colors. Thus quarks at the same point in space were no longer identical and no longer violated the Fermi statistics.

Since 1966 there have been great advances in our understanding of the forces between quarks as a result of the work of many physicists all over the world. There have also been new experimental discoveries. We now have a theory called quantum chromodynamics (QCD) which is a beautiful generalization of the well-known theory of electricity and magnetism to describe the strong forces between quarks. The new field of force analogous to the electromagnetic field is called the gluon field, because it provides the "glue" which holds quarks together in protons. There are two kinds of forces between quarks, called "chromoelectric" and "chromomagnetic," directly analogous to the ordinary electric and magnetic forces between electrons and nuclei in atoms. There has also been the experimental discovery of new very heavy quarks of a new fourth flavor called "charmed" quarks found in particles discovered since 1974. The QCD theory tells us that quarks of all flavors have the same chromoelectric charge and chromoelectric forces, whether they are up, down, strange, or charmed, just as the ordinary electric force between two protons is the same as the ordinary electric force between two electrons. But the chromomagnetic force is weaker for heavier than for lighter quarks, just as the ordinary magnetic force on a proton is 1000 times weaker than the ordinary magnetic force on the much lighter electron. The force also depends upon the orientation of the quarks, which behave like tiny magnets. They can either attract or repel one another depending upon which way they are pointing.

In 1975 Sakharov extended the application of his old 1966 mass formula to the newly discovered particles containing charmed quarks. In 1980, sitting in Gor'kii, he improved the model with the aid of these new ideas from QCD. At the same time, I did a similar updating of our 1966 work at the Weizmann Institute and obtained essentially the same new results as Sakharov. The old formula gave the mass of any particle as the sum of the masses of its constituent quarks with added contributions from the energies of the unknown interquark forces. Now we can use the additional information that these quark forces are known from QCD to be chromoelectric and chromomagnetic. The chromoelectric force is the same for all quarks and does not depend upon flavor or orientation. The chromomagnetic force is inversely proportional to the quark masses and depends in a well-defined way on the orientations of the quarks. These newly understood properties of the forces lead to new relations between the masses of different particles.

There are pairs of particles believed to be made of exactly the same quarks, and to differ only by having their "chromomagnets" pointing in different directions. The masses of such a pair of particles differ only by the change in energy of the chromomagnetic force. We can now compare this chromomagnetic energy for particles made of different combinations of light quarks, heavier strange quarks, and very heavy charmed quarks. Simple predictions for the differences in mass between such pairs of particles are then obtained from the theory that the chromomagnetic forces are inversely proportional to the masses of the quarks. These predictions are found to be in excellent agreement with experiment.

We can also compare the masses of pairs of particles which differ only by having a light quark replaced by a strange or charmed quark. The proton consists of three light quarks. A particle called the lambda has two light quarks and one strange quark. The lambda is heavier than the proton by just the difference between the mass of the strange quark and the mass of the light quark. There are also particles called mesons, made of two quarks. Some mesons are made of two light quarks, and some heavier "strange" mesons are made of one light quark and one strange quark. The strange mesons are heavier than the ordinary mesons also by just the mass difference between strange and light quarks. If this model is correct the strange mesons must be heavier than the ordinary mesons by exactly the same amount as the lambda is heavier than the proton. This is found experimentally to be true, thus confirming the hypothesis that mesons and protons are made of the same quarks.

Thus, even though nobody has succeeded in breaking up the proton into quarks and showing that isolated quarks exist, there are many experimental tests of the theory that particles are made of quarks bound together by chromoelectric and chromomagnetic forces, and that the strengths of these forces depend upon the masses and orientations of the quarks in the way suggested by the QCD theory. Sakharov has also calculated the strengths of these QCD forces from the magnitudes of the observed mass differences.

We now examine the technical details of Sakharov's calculations.

Compare the proton made of three light quarks and the Λ made from two light quarks and one heavier quark. The Λ is heavier than the proton, and we can use the measured values of the masses to estimate the mass difference between the heavier quark and the light quark.

$$M(\Lambda) - M(\text{proton}) = m(\text{heavier quark}) - m(\text{light quark}) \quad (1a)$$

Particles made of three quarks are called baryons. There are also particles called mesons made of two quarks. The π and ρ mesons are made of two light quarks and the K and K^* mesons are made of one light quark and one heavier quark. We can also estimate the mass difference between the heavier and the light quark from these meson masses:

$$\tfrac{3}{4}[m(K^*) - m(\rho)] + \tfrac{1}{4}[m(K) - m(\pi)] = m(\text{heavier}) - m(\text{light}) \quad (1b)$$

The numbers $\tfrac{3}{4}$ and $\tfrac{1}{4}$ arise in order to take proper account of the chromomagnetic forces.

These two estimates are in remarkable agreement. The first gives 177 MeV; the second 179 MeV.

In the same way we obtain relations between masses of other particles, including the particles D, D^*, and Λ_c containing the very heavy charmed quarks:

$$M(\Lambda_c) - M(\text{proton}) = m(\text{very heavy}) - m(\text{light}) \quad (2a)$$

$$\tfrac{3}{4}[M(D^*) - M(\rho)] + \tfrac{1}{4}[M(D) - M(\pi)] = m(\text{very heavy}) - m(\text{light}) \quad (2b)$$

This is also in good agreement with the experimental values.

We now examine pairs of particles like (Δ, N), (ρ, π), (K^*, K), (Σ, Λ), (D^*, D) and (Σ_c, Λ_c). Both particles in each pair are made of the same quarks with their chromomagnets pointing in different directions. Their mass difference gives the chromomagnetic energy. Using the QCD ideas of flavor dependence of these chromomagnetic energies we obtain additional mass relations:

$$\frac{M(K^*) - M(K)}{M(\rho) - M(\pi)} + \frac{3}{2} \frac{M(\Sigma) - M(\Lambda)}{M(\Delta) - M(N)} = 1 \quad (\text{experimentally } 1.02) \quad (3a)$$

$$\frac{M(D^*) - M(D)}{M(\rho) - M(\pi)} + \frac{3}{2} \frac{M(\Sigma_c) - M(\Lambda_c)}{\Delta - N} = 1 \quad (\text{experimentally } 1.08) \quad (3b)$$

$$\frac{M(\Delta) - M(N)}{M(\rho) - M(\pi)} = \frac{1}{2} \quad \text{(experimentally 0.46)} \tag{4}$$

We now use the experimentally determined masses of particles to estimate the strength of the force between the quarks and the gluon field. There are two very similar light quarks called u and d. The d quark is just a little heavier than the u quark and has a different electric charge. When we change a u quark to a d quark in a particle, the ordinary electromagnetic forces change because the charge has changed, and the chromomagnetic forces also change. By comparing the change in mass due to ordinary electromagnetic forces with the change in mass due to chromomagnetic forces we can relate the strength of the new chromomagnetic force to the well-known electromagnetic forces.

The mass difference between two particles which differ by having a u quark changed into a d quark depends on two quantities, the ratio e^2/g^2 of the electric force to the chromoelectric force and the ratio m_d/m_u of the mass of the d quark to the mass of the u quark. Following our same approach we find

$$\frac{M(\rho^0) - M(\pi^0)}{M(\rho^+) - M(\pi^+)} = 1 + \frac{e^2}{2g^2} \tag{5a}$$

$$\frac{M(K^{*0}) - M(K^0)}{M(K^{*+}) - M(K^+)} = 2 + \frac{e^2}{3g^2} - \frac{m_d}{m_u} \tag{5b}$$

$$\frac{M(D^{*0}) - M(D^0)}{M(D^{*+}) - M(D^+)} = \frac{2e^2}{3g^2} + \frac{m_d}{m_u} = \frac{M(\Xi^{*0}) - M(\Xi^0)}{M(\Xi^{*-}) - M(\Xi^-)} \tag{6a}$$

$$\frac{M(\Sigma^{*+}) - M(\Sigma^+)}{M(\Sigma^{*-}) - M(\Sigma^-)} = \frac{2e^2}{3g^2} + \frac{m_d}{m_u} = \frac{M(\Xi^{*0}) - M(\Xi^0)}{M(\Xi^{*-}) - M(\Xi^-)} \tag{6b}$$

The last two of these relations are in good agreement with experiment. They can be combined with the first two to give values for e^2/g^2. Using the known values of e^2, we obtain

$$\frac{g^2}{4\pi} = 0.255^{+0.081}_{-0.049} \quad \text{from the } \rho \text{ and } \pi \tag{7}$$

$$\frac{g^2}{4\pi} = 0.369^{+0.16}_{-0.08} \quad \text{from the } D, K, \text{ and } \Sigma \tag{8}$$

A detailed derivation of these results is obtained from the following general mass formulas for three-quark baryons and two-quark mesons:

$$M = \delta_M + m_1 + m_2 + b \frac{m_0}{m_1} \frac{m_0}{m_2} (\boldsymbol{\sigma}_1 \cdot \boldsymbol{\sigma}_2) \tag{9}$$

$$M = \delta_B + m_1 + m_2 + m_3$$
$$+ \frac{b}{3} \left[\frac{m_0}{m_1} \frac{m_0}{m_2} (\boldsymbol{\sigma}_1 \cdot \boldsymbol{\sigma}_2) + \frac{m_0}{m_2} \frac{m_0}{m_3} (\boldsymbol{\sigma}_2 \cdot \boldsymbol{\sigma}_3) + \frac{m_0}{m_3} \frac{m_0}{m_1} (\boldsymbol{\sigma}_3 \cdot \boldsymbol{\sigma}_1) \right] \tag{10}$$

Here δ_M, δ_B, b are parameters with the dimension of mass, and m_i are masses of quarks; m_0 is the mass of an "ordinary" quark u or d; m_s and m_c are the masses of the strange and charmed quark, respectively. Thus we have six parameters.

The last term in (9) and (10) is the spin-spin interaction of quarks, and $\boldsymbol{\sigma}_i \cdot \boldsymbol{\sigma}_j$ is the scalar product of quark spin vectors. The underlying physics in these formulas is the following.

The chromoelectric contribution to the mass is independent of flavor and has a universal value for all mesons and a universal value for all baryons. These are included in the parameters δ_M and δ_B. The chromomagnetic contribution to the masses is given by a first-order hyperfine interaction proportional to the spin orientation factor $(\boldsymbol{\sigma}_i \cdot \boldsymbol{\sigma}_j)$ and inversely proportional to the quark mass production $m_i m_j$. The only dependence on flavor is this quark mass factor. Thus the strength of this interaction is given by a universal parameter b. There remains the question of how the chromomagnetic forces in mesons and baryons are related. QCD gives a unique answer for the strengths of the forces, but their contribution to the masses also depends upon the size of the system. Because baryons are larger than mesons, the interacting quarks in mesons are closer together than in the larger baryons, and the forces have a stronger effect. Sakharov takes this into account with factor $b/3$ in the baryon formula and b in the meson formula. This factor $\frac{1}{3}$ includes a factor $\frac{1}{2}$ from QCD and an additional factor $\frac{2}{3}$ for the difference in size. In my paper I use a slightly different argument for the effects of the difference in size of the mesons and baryons and obtain $\frac{3}{4}$ instead of $\frac{2}{3}$ for the size factor and $3b/8$ instead of $b/3$ in the baryon mass formula. The difference between these two values has only a small effect on the results. Equations (1) to (3) do not depend upon this factor at all, and the right-hand side of Eq. (4) is $\frac{9}{16}$ in my paper.

I conclude with an interesting piece of "forgotten ancient history" which has interesting implications today. In their 1966 paper Sakharov and Zel'dovich point out that their successful relation (3a) differs by a factor $\frac{3}{2}$ from an $SU(6)$ relation which disagrees with experiment. I looked up the $SU(6)$ reference and was amazed to find H. Harari and H. J. Lipkin, *Phys. Rev. Lett.* 13:345 (1964); a paper that I had forgotten completely. In this $SU(6)$ paper we attempted to generalize the symmetry approach to hadron masses which had proved so successful in $SU(3)$ with the Gell-Mann–Okubo mass formula. We assumed that the $SU(6)$ symmetry breaking in the mass spectrum transformed in a very definite way under $SU(6)$ and found that it was impossible to get agreement with the observed masses. The $SU(6)$ symmetry-breaking operators required to fit meson and baryon masses were different. The mass formulas (9) and (10) explain this difference between $SU(3)$ and $SU(6)$. The flavor-dependent terms in both formulas transform under $SU(3)$ like the isoscalar member of an octet to a very good approximation. But the $SU(6)$ transformation properties of (9) and (10) are complicated and different.

In 1965 the underlying physics behind the successful $SU(6)$ classification of hadrons was very unclear. The symmetry approach was widely used, with attempts to embed $SU(6)$ in some larger group including both space-time and internal symmetries. These failed because the underlying basis for the $SU(6)$ classification was not a higher symmetry but the composite nature of the hadrons, as Sakharov and Zel'dovich already realized in 1966. Today we are back at the same problem on a deeper level. Instead of unifying mesons and baryons, we are looking for a "grand unification" scheme including both quarks and leptons. Both the symmetry and composite-model approaches are being investigated. It is still too early to see whether the quark-lepton spectrum arises from representations of a new grand-unification gauge group or from a common composite structure based on new building blocks like rishons. It would be very useful today to find crucial clues in the experimental data that would distinguish between the two approaches, like the factor $\frac{3}{2}$ found by Sakharov and Zel'dovich in 1966.

REFERENCES

The 1966 papers on the quark structure and masses of strongly interacting particles:
1. Ya. B. Zel'dovich and A. D. Sakharov, *J. Nuclear Phys.* 4:395 (1966); *Sov. J. of Nuclear Phys.* 4:283 (1967), trans. [Paper 19 in this volume].
2. P. Federman, H. R. Rubinstein, and I. Talmi, *Phys. Lett.* 22:203 (1966); H. R. Rubinstein, *Phys. Lett.* 22:210 (1966).

The introduction of color:
3. O. W. Greenberg, *Phys. Rev. Lett.* 13:598 (1964).

Extension to include charm:
4. A. D. Sakharov, *ZhETF Pis'ma* 21:554 (1975); *JETP Lett.* 21:258 (1975), trans. [Paper 20 in this volume].

Introduction of QCD ideas to hadron masses (flavor-independent chromoelectric and mass-dependent chromomagnetic interactions):

5. A. DeRujula, H. Georgi, and S. L. Glashow, *Phys. Rev. D12*:147 (1975).

Use of QCD ideas to update 1966 mass formula:

6. H. J. Lipkin, *Phys. Lett. 74B*:399 (1978).
7. A. D. Sakharov, *ZhETF 78*:2112 (1980); *Sov. Phys. JETP 51*:1059 (1980), trans. [Paper 21 in this volume].

Comparison of chromomagnetic and electromagnetic interactions in states with u and d quarks, Eqs. (5) to (6):

8. I. Cohen and H. J. Lipkin, *Phys. Lett. 84B*:323 (1979).
9. A. D. Sakharov, *ZhETF 79*:350 (1980); *Sov. Phys. JETP 52*:175 (1980), trans. [Paper 22 in this volume].

A general review of the situation in 1978 is given in

10. H. J. Lipkin, in *Common Problems in Low- and Medium-Energy Nuclear Physics*, (NATO ASI Series B, Physics, Volume 45), the Proceedings of NATO Advanced Study Institute/1978 Banff Summer Institute on Nuclear Theory held at Banff, Canada from August 21 to September 1, 1978, (B. Castel, B. Goulard, and F. C. Khanna, eds.), Plenum Publishing Co., New York, p. 175 (1979).

Part 4

DIVERTISSEMENTS

Divertissements

1. A cloud of rarefied gas with equation of state $p \sim \rho T$ is in a radiation field in thermal equilibrium with the radiation; the mean free path of the radiation is much greater than the cloud diameter and the temperature of the radiation is a function of the time. Find a self-similar solution for the expansion of the cloud. The answer is

$$\rho \sim \frac{1}{r_0^3} \exp\left(-\frac{r^2}{r_0^2}\right)$$

where r_0 is a function of the time.

2. A jet of viscous fluid exhausts from a circular opening (radius r_0) and is stretched under the influence of gravity. Find the shape of the jet at a distance l from the opening ($l \gg r_0$). We ignore surface tension and inertia. The answer is $r \sim l^{-1}$.

3. On the flat interface of two transparent media there is a light-absorbing pigment. At $t = 0$, a beam of light is incident on the interface in the form of a small disk. Find the law of increase of the temperature at the center of the disk. We find the exact solution by integrating the Green's function of the heat conduction equation. At the initial period, the solution can be approximated by the solution to the one-dimensional problem and $\Delta T \sim \sqrt{t}$; the distribution with respect to the coordinate is self-similar and described by the solution of Poisson [equation] with an initial distribution of temperature

$$T = \begin{cases} \dfrac{2j}{k}|x| & \text{for } x \leq 0 \\ 0 & \text{for } x \geq 0 \end{cases}$$

(the domain of the solution), where j is a given flow of heat at the point

$x = 0$, and k is a coefficient of thermal conductivity (this limit is, of course, well known).

4. Find the force of electrostatic attraction of two convex conducting bodies when the minimal distance Δ between them is much less than the radius of curvature (for example, two cylinders of radius $R \gg \Delta$ with axes inclined at angle α). The potential difference V between the bodies is given. The answer is $f = V^2R/(4\Delta \sin \alpha)$. I came across this problem in connection with an analogous problem in the theory of magnetism. During work in a factory during the Second World War, I proposed a simple method for determining the thickness of nonmagnetic coatings of bullets in a geometry analogous to the one for which the electrostatic problem is solved. (At that time, 1943, I also proposed a magnetic device for detecting nontempered centers of armor-piercing projectiles. This proposal made it unnecessary to carry out sample testing of a large number of armor-piercing projectiles by breaking them. I took out an inventor's certificate in 1944.)

5. When cabbage is cut up with a knife, one obtains polygons with different numbers of vertices and different sizes and shapes. Determine the mean number of vertices $\bar{n}$ and the ratio of the square of the mean perimeter $\bar{L}$ to the mean area $\bar{S}$. The answer is $\bar{n} = 4$ and $\bar{L}^2/\bar{S} = 4\pi$ (i.e., the same as for a circle, which appears remarkable at first glance). I was led to consider this problem when cutting cabbage to help my wife make pies.

6. N points on a plane are given. Each point is connected by a colored line to each of the remaining $N - 1$ points, and p colors are used. Find a function $L(n,p)$ such that: (1) For $N > L(n,p)$ and any choice of the colors for the $N(N - 1)/2$ lines there are at least n points such that all the $n(n - 1)/2$ lines joining them are of the same color. (2) For $N < L(n,p)$, one can choose the colors of the lines in such a way that no n points are joined by lines of one color.[†] The complete answer was found only for some cases. Probably $L(3,p) = [ep!]$. In this case 1 holds, and 2 holds for $p = 2,3$. For $n = 3$, $L' > L$ were found such that for $N > L'$ assertion 1 is true.

7. Liquid is poured into a circular vessel which stands on a table. Ink spots are deposited on the surface of the liquid. The vessel is turned by hand through some angle ϕ, the law of increase and decrease in the

---
[†]Statement 1 is called the Ramsey theorem in finite (and infinite) combinatorics. The number $L(n,p)$ is called, correspondingly, a Ramsey number, and its determination is one of the difficult combinatorial problems, cf., e.g., M. Hall, *A Survey of Combinatorial Analysis*, John Wiley & Sons, New York, 1958; R. L. Graham, *Rudiments of Ramsey Theory*, AMS, No. 45, 1981.—EDS.

angular velocity being arbitrary, provided laminar flow of the concentric layers is guaranteed. Prove that after the motion of the liquid stops the configuration of spots, turned through angle ϕ, is recovered. Two solutions to the problem have been found, one by myself and the other by E. I. Zababakhin. My solution is based on the use of a Fourier integral. We Fourier transform the function $\chi(t,r) = (d/dt)\phi(t,r)$;

$$\chi_\varepsilon(\varepsilon,r) = \frac{1}{(2\pi)^{1/2}} \int \exp(i\varepsilon t)\chi(t,r)\, dt$$

does not depend on r as $\varepsilon \to 0$. Hence, $\Delta\phi$ does not depend on r.

8. Two families of theorems relating to number theory have been formulated. (1) The sequences $a_n = n! + 1$, $a_n = [n \ln(\ln n)]! + 1$, etc. (here the square brackets denote the integral part), contain an infinite number of primes, since the series $\Sigma(1/\ln a_n)$ diverges. (2) The sequences

$$b_n = (n^2)! + 1 \qquad b_n = [n(\ln n)\alpha]! + 1 \qquad \alpha > 0$$

for which $\Sigma(1/\ln b_n)$ converges, contain a finite number of primes.

9. Using the identity

$$\sqrt{a_0^2 - 1/b_0^2} = \lim \left(\frac{a_n}{b_n}\right) \qquad \text{where } a_n = 2a_{n-1}^2 - 1$$
$$\text{and } b_n = 2a_{n-1}b_{n-1}$$

I have constructed a rapidly converging algorithm for calculating the square roots of all integers, and an algorithm for calculating the terms f_{2n} and f_{2n+1} of the Fibonacci series if the members f_n and f_{n+1} are known (it is not necessary to calculate all the intermediate terms).

10. I have devised some simple approximate constructions for solving the problem of trisecting an angle ψ. The simplest of them is as follows. One constructs an isosceles triangle with angle ψ at the vertex A and base a. From the vertex B one marks out the interval $BE = (a/3)(1 - \sqrt{2/3}) \approx 11a/180$ and the interval $ED = a/3$. Then the angle $EAD \approx \psi/3$.

11. A proposal for a simple example of hydrodynamic motion leading to a "magnetohydrodynamic dynamo." In an incompressible conducting medium we consider a closed torus with "frozen" magnetic field H, large radius R, section S, and magnetic flux SH (Fig. 1a). The torus is stretched by a factor two:

$$R \to 2R \qquad S \to \frac{S}{2} \qquad H \to 2H$$

Then the field energy $H^2SR/4$ increases by four times (Fig. 1b). The torus is then twisted into a figure eight without changing the section or length (Fig. 1c). Then the figure eight is folded in half to make a torus of the original dimensions but with twice the field (Fig. 1d). The process can be repeated any number of times.

12. *Icing* A big flat piece of ice with an initial temperature of $T_0 < 0°C$ is submerged in water of temperature $T = 0°C$. Find the law of growth of the thickness of the ice (one-dimensional problem). Answer: The distribution of temperature is given by the Poisson solution with the following initial distribution:

$$T = \begin{cases} T_0 < 0 & \text{for } x < 0 \\ T_1 > 0, T_1 < |T_0| & \text{for } x > 0 \end{cases}$$

The thickness of the ice mound is $\Delta = \alpha t^{1/2}$. The coefficient α and parameter T_1 are determined from the system of equations (using the self-similarity of the problem, time is excluded from this system):

$$T(\Delta) = 0$$

$$K \frac{\partial T}{\partial x}\bigg|_\Delta = Q \frac{d\Delta}{dt}$$

Here $K = Ca$ is the heat conduction, C is heat capacity, a is temperature conductivity, Q is latent heat of fusion. For $|T_0| \ll Q/C$ we take $T_1 = |T_0|$ in the first approximation. Next we find $\alpha = 2|T_0|a^{1/2}C/(Q\pi^{1/2})$ from the second equation, and then, from the first equation we find the difference $|T_0| - T_1 \sim T_0^2$.

I solved this problem in 1943, when I was working at the factory and was stimulated by reading some book on metallurgy. The problem is an abstract version of more complex thermic processes during phase transformations of steel. Probably, this is the beginning of my work as a theoretical physicist. It was never published and was not reported anywhere.

13. *Textbook and popular scientific papers*

1. In collaboration with M. I. Bludov, I have been preparing new editions of the physics textbook for technical schools written by my father D. I. Sakharov together with Bludov. In 1963 (or 1964) the first revised edition was published. The two final chapters were written by me: "Optical quantum effects" and "The atomic nucleus." In 1974, a new revised edition, in which my part was greater, should have appeared, but the permission for publication was withdrawn after the newspaper campaign against me in 1973. I have prepared for reissue my father's book *Exercise Problems in Physics* (last edition, 1973).

2. A paper in the collection *The Future of Science* edited by

V. A. Kirillin (1967). The book was not put on sale. My paper contains predictions about the development of science and technology. Some ideas from the paper subsequently appeared in my book *Progress, Coexistence, and Intellectual Freedom* (W. W. Norton, New York, 1968).
3. A paper entitled "Symmetry of the universe," in the collection *The Future of Science* published by *Znanie* (Knowledge) *Publishing House* in 1967. This is an attempt to give a popular account of C, P, CP, and CPT symmetry and my ideas about cosmological CPT symmetry.
4. A paper in 1969 in the journal *Fizika v Shkole* (Physics in Schools): "Does an elementary length exist?" This is an attempt at a popular exposition of some of the ideas of local and nonlocal field theory.

These figures are reproduced directly from Sakharov's drawings.

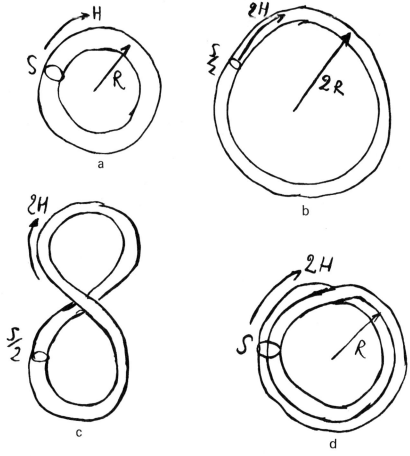

Figure 1

Part 5

NOBEL PEACE PRIZE LECTURE OF 1975

The Nobel Prize Lecture, 1975

Peace, Progress, and Human Rights

Citation for the 1975 Nobel Peace Prize Award

The Nobel Committee of the Norwegian Parliament has awarded Nobel's peace prize for 1975 to Andrei Sakharov.

Sakharov's personal and fearless effort in the cause of peace among mankind serves as a mighty inspiration to all true endeavors to promote peace. Uncompromisingly and forcefully, Sakharov has fought not only against the abuse of power and violations of human dignity in all its forms, but he has with equal vigor fought for the ideal of a state founded on the principle of justice for all.

In a convincing fashion Sakharov has emphasized that the individual rights of man can serve as the only sure foundation for a genuine and long-lasting system of international cooperation. In this manner he has succeeded very effectively, and under trying conditions, in reinforcing respect for such values as all true friends of peace are anxious to support.

Andrei Dmitrivich Sakharov has addressed his message of peace and justice to all peoples of the world. For him it is a fundamental principle that world peace can have no lasting value, unless it is founded on respect for the individual human being in society. This respect has found expression in several international declarations; for example, the UN declaration on the rights of man. Sakharov has demanded that the national authorities of each country must live up to the commitments they have undertaken in signing these declarations.

In the various agreements signed this year by 35 states at the security conference in Helsinki, it was again emphasized that this respect for human dignity was an obligation undertaken by the states themselves. In the agreement the parties acknowledge that respect for human rights and fundamental freedoms is an important factor in the cause of peace, justice, and well-being which is essential to ensure the development of friendly relations and cooperation not only among themselves but among all the countries of the world.

In more forceful terms than others, Andrei Sakharov has warned us against not taking this seriously, and he has placed himself in the vanguard of the efforts

Copyright © 1975 The Nobel Foundation. Reprinted with permission.

to make the ideals expressed in this paragraph of the Helsinki agreement a living reality.

Andrei Sakharov is a firm believer in the brotherhood of man, in genuine coexistence, as the only way to save mankind. It was precisely by means of encouraging fraternization between all peoples, based on truth and sincerity, that Alfred Nobel envisaged the possibilities of creating a safer future for all mankind. When states violate the fundamental precepts of human rights, they are also, in Sakharov's view, undermining the work to promote confidence across national borders.

Sakharov has warned against the dangers connected with a bogus detente based on wishful thinking and illusions. As a nuclear physicist he has, with his special insight and sense of responsibility, been able to speak out against the dangers inherent in the armaments race between the states. His aims are demilitarization, democratization of society in all countries and a more rapid pace of social progress.

Sakharov's love of truth and strong belief in the inviolability of the human being, his fight against violence and brutality, his courageous defense of the freedom of the spirit, his unselfishness and strong humanitarian convictions have turned him into the spokesman for the conscience of mankind, which the world so sorely needs today.

Peace, progress, human rights—these three goals are indissolubly linked: it is impossible to achieve one of them if the others are ignored. This idea provides the main theme of my lecture.

I am deeply grateful that this great and significant award, the Nobel Peace Prize, has been given to me, and that I have the opportunity of addressing you here today. I was particularly gratified at the Committee's citation, which stresses the defense of human rights as the only sure basis for genuine and lasting international cooperation. This idea is very important to me; I am convinced that international trust, mutual understanding, disarmament, and international security are inconceivable without an open society with freedom of information, freedom of conscience, the right to publish, and the right to travel and choose the country in which one wishes to live. I am also convinced that freedom of conscience, together with other civic rights, provides both the basis for scientific progress and a guarantee against its misuse to harm mankind, as well as the basis for economic and social progress, which in turn is a political guarantee making the effective defense of social rights possible. At the same time I should like to defend the thesis of the original and decisive significance of civic and political rights in shaping the destiny of mankind. This view differs essentially from the usual Marxist theory, as well as from technocratic opinions, according to which only material factors and social and economic conditions are of decisive importance. (But in saying this, of course, I have no intention of denying the importance of people's material welfare.)

I should like to express all these theses in my lecture, and in particular to dwell on a number of specific problems affecting the violation of human rights. A solution of these problems is imperative, and the time at our disposal is short.

This is the reason why I have called my lecture "Peace, Progress, and Human Rights." There is, naturally, a conscious parallel with the title of my 1968 article "Thoughts on Progress, Peaceful Coexistence, and Intellectual Freedom," with which my lecture, both in its contents and its implications, has very close affinities.

There is a great deal to suggest that mankind, at the threshold of the second half of the twentieth century, entered a particularly decisive and critical historical era.

Nuclear missiles exist capable in principle of annihilating the whole of mankind; this is the greatest danger threatening our age. Thanks to economic, industrial, and scientific advances, so-called "conventional" arms have likewise grown incomparably more dangerous, not to mention chemical and bacteriological instruments of war.

There is no doubt that industrial and technological progress is the most important factor in overcoming poverty, famine, and disease. But this progress leads at the same time to ominous changes in the environment in which we live and to the exhaustion of our natural resources. Thus, mankind faces grave ecological dangers.

Rapid changes in traditional forms of life have resulted in an unchecked demographic explosion which is particularly noticeable in the developing countries of the Third World. The growth in population has already created exceptionally complicated economic, social, and psychological problems and will in the future inevitably pose still more serious problems. In many countries, particularly in Asia, Africa, and Latin America, the lack of food will be an overriding factor in the lives of many hundreds of millions of people, who from the moment of birth are condemned to a wretched existence on the starvation level. Moreover, future prospects are menacing, and in the opinion of many specialists, tragic, despite the undoubted success of the "green revolution."

But even in the developed countries, people face serious problems. These include the pressure resulting from excessive urbanization, all the changes that disrupt the community's social and psychological stability, the incessant pursuit of fashion and trends, overproduction, the frantic, furious tempo of life, the increase in nervous and mental disorders, the growing number of people deprived of contact with nature and of normal human lives, the dissolution of the family and the loss of simple human pleasures, the decay of the community's moral and ethical principles, and the loss of faith in the purpose of life. Against this background there is a whole host of ugly phenomena: an increase in crime, in alcoholism, in drug addiction, in terrorism, and so forth. The imminent exhaustion of the world's resources, the threat of overpopulation, the constant and deep-rooted international, political, and social problems are making a more and more forceful impact on the developed countries too, and will deprive—or at any rate threaten to deprive—a great many people who are accustomed to abundance, affluence, and creature comforts.

However, in the pattern of problems facing the world today a more decisive and important role is played by the global political polarization of mankind, which is divided into the so-called First World (conventionally called the Western world), the Second (socialist), and the Third (the developing countries). Two powerful socialist states, in fact, have become mutually hostile totalitarian empires, in which a single party and the state exercise immoderate power in all spheres of life. They possess an enormous potential for expansion, striving to increase their influence to cover large areas of the globe. One of these states—the Chinese People's Republic—has reached only a relatively modest stage of economic development, whereas the other—the Soviet Union—by exploiting its unique natural resources, and by taxing to the utmost the powers of its inhabitants and their ability to suffer continued privation, has built up a tremendous war potential and a relatively high, though one-sided, economic development. But in the Soviet Union, too, the people's standard of living is low, and civic rights are more restricted than in less socialist countries. Highly complicated global problems also affect the Third World, where relative economic stagnation goes hand in hand with growing international political activity.

Moreover, this polarization further reinforces the serious dangers of nuclear annihilation, famine, pollution of the environment, exhaustion of resources, overpopulation, and dehumanization.

If we consider this complex of urgent problems and contradictions, the first point that must be made is that any attempt to reduce the tempo of scientific and technological progress, to reverse the process of urbanization, to call for isolationism, patriarchal ways of life, and a renaissance based on ancient national traditions, would be unrealistic. Progress is indispensable, and to halt it would lead to the decline and fall of our civilization.

Not long ago we were unfamiliar with artificial fertilizers, mechanized farming, chemical pesticides, and intensive agricultural methods.

There are voices calling for a return to more traditional and possibly less dangerous forms of agriculture. But can this be accomplished in a world in which hundreds of millions of people are suffering from hunger? On the contrary, there is no doubt that we need increasingly intensive methods of farming, and we must spread modern methods all over the world, including the developing countries.

We cannot reject the idea of a spreading use of the results of medical research or the extension of research in all its branches, including bacteriology and virology, neurophysiology, human genetics, and gene surgery, no matter what potential dangers lurk in their abuse and the undesirable social consequences of this research. This also applies to research in the creation of artificial intelligence systems, research involving behavior, and the establishment of a unified system of global communication, systems for selecting and storing information, and so forth. It is quite clear that in the hands of irresponsible bureaucratic authorities operating secretly, all this research may prove exceptionally dangerous, but at the same time it may prove extremely important and necessary to mankind, if it is carried out under public supervision and discussion and socioscientific analysis. We cannot reject wider application of artificial materials, synthetic food, or the modernization of every aspect of life; we cannot obstruct growing automation and industrial expansion, irrespective of the social problems these may involve.

We cannot condemn the construction of bigger nuclear power stations or research into nuclear physics, since energetics is one of the bases of our civilization. In this connection I should like to remind you that 25 years ago I and my teacher, the winner of the Nobel Prize for Physics, Igor Evgenyevich Tamm, laid the basis for nuclear research in our country. This research has achieved tremendous scope, extending into the most varied directions, from the classical method for magnetic heat insulation to those for the use of lasers.

We cannot cease interplanetary and intergalactic space research, including the attempts to intercept signals from civilizations outside our own earth. The chance that such experiments will prove successful is probably small, but precisely for this reason the results may well be tremendous.

I have mentioned only a few examples. In actual fact all important aspects of progress are closely interwoven; none of them can be discarded without the risk of destroying the entire structure of our civilization. Progress is indivisible. But intellectual factors play a special role in the mechanism of progress. Underestimating these factors is particularly widespread in the socialist countries, probably due to the populist-ideological dogmas of official philosophy, and may well result in distortion of the path of progress or even its cessation and stagnation.

Progress is possible and innocuous only when it is subject to the control of reason. The important problems involving environmental protection exemplify the role of public opinion, the open society, and freedom of conscience. The partial liberalization in our country after the death of Stalin made it possible to engage in public debate on this problem during the early 1960s. But an effective solution demands increased tightening of social and international control. The military application of scientific results and controlled disarmament are an equally critical area, in which international confidence depends on public opinion and an open society. The example I gave involving the manipulation of mass psychology is already highly topical, even though it may appear farfetched.

Freedom of conscience, the existence of an informed public opinion, a pluralistic system of education, freedom of the press, and access to other sources of information—all these are in very short supply in the socialist countries. This situation is a result of the economic, political, and ideological monism which is characteristic of these nations. At the same time these conditions are a vital necessity, not only to avoid all witting or unwitting abuse of progress, but also to strengthen it.

An effective system of education and a creative sense of heredity from one generation to another are possible only in an atmosphere of intellectual freedom. Conversely, intellectual bondage, the power and conformism of a pitiful bureaucracy, acts from the very start as a blight on humanistic fields of knowledge, literature, and art, and results eventually in a general intellectual decline, the bureaucratization and formalization of the entire system of education, the decline of scientific research, the thwarting of all incentive to creative work, stagnation, and dissolution.

In the polarized world the totalitarian states, thanks to détente, today may indulge in a special form of intellectual parasitism. And it seems that if the inner changes that we all consider necessary do not take place, those nations will soon be forced to adopt an approach of this kind. If this happens, the danger of an explosion in the world situation will merely increase. Cooperation between the Western states, the socialist nations, and the developing countries is a vital necessity for peace, and it involves exchanges of scientific achievements, technology, trade, and mutual economic aid, particularly where food is concerned. But this cooperation must be based on mutual trust between open societies, or—to put it another way—with an open mind, on the basis of genuine equality and not on the basis of the democratic countries' fear of their totalitarian neighbors. If that were the case, cooperation would merely involve an attempt at ingratiating oneself with a formidable neighbor. But such a policy would merely postpone the evil day, soon to arrive anyway and, then, ten times worse. This is simply another version of Munich. Détente can only be assured if from the very outset it goes hand in hand with continuous openness on the part of all countries, an aroused sense of public opinion, free exchange of information, and absolute respect in all countries for civic and political rights. In short: in addition to détente in the material sphere, with disarmament and trade, détente should take place in the intellectual and ideological sphere. President Giscard d'Estaing of France expressed himself in an admirable fashion during his visit to Moscow. Indeed, it was worth enduring criticism from shortsighted pragmatists among his countrymen to support such an important principle.

Before dealing with the problem of disarmament I should like to take this opportunity to remind you of some of my proposals of a general nature. First and foremost is the idea of setting up an international consultative committee for questions related to disarmament, human rights, and the protection of the environment, under the aegis of the United Nations. In my opinion a committee of this kind should have the right to exact replies from all governments to its inquiries and recommendations. The committee could become an important working body in securing international discussion and information on the most important problems affecting the future of mankind. I hope this idea will receive support and be discussed.

I should also emphasize that I consider it particularly important for United Nations armed forces to be used more generally for the purpose of restricting armed conflicts between states and ethnic groups. I have a high regard for the United Nations role, and I consider the institution to be one of mankind's most important hopes for a better future. Recent years have proved difficult and critical for this organization. I have written on this subject in *My Country and the World*, but after it was published, a deplorable event took place: The General Assembly adopted—without any real debate—a resolution declaring Zionism a form of racism and racial discrimination. Zionism is the ideology of a national rebirth of the Jewish people after 2000 years of diaspora, and it is not directed against any other people. The adoption of a resolution of this kind has damaged the prestige of the United Nations. But despite such motions, which are frequently the result of an insufficient sense of responsibility among leaders of some of the UN's younger members, I believe nevertheless that the organization may sooner or later be in a position to play a worthy role in the life of mankind, in accordance with its Charter's aims.

Let me now address one of the central questions of the present age, the problem of disarmament. I have described in detail just what my position is in *My Country and the World*. It is imperative to promote confidence between nations, and carry out measures of control with the aid of international inspection groups. This is only possible if détente is extended to the ideological sphere, and it presupposes greater openness in public life. I have stressed the need for international agreements to limit arms supplies to other states, special agreements to halt production of new weapons systems, treaties banning secret rearmament, the elimination of strategically unbalancing factors, and in particular a ban on multiwarhead nuclear missiles.

What would be the ideal international agreement on disarmament on the technical plane?

I believe that prior to such an agreement we must have an official declaration—though not necessarily public in the initial stages—on the extent of military potential (ranging from the number of nuclear warheads to forecast figures on the number of personnel liable for military service), with, for example, an indication of areas of "potential confrontation." The first step would be to ensure that for every single strategic area and for all sorts of military strength an adjustment would be made to iron out the superiority of one party to the agreement in relation to the other. (Naturally this is the kind of pattern that would be liable to some adjustment.) This would in the first place obviate the possibility of an agreement in one strategic area (Europe, for instance) being utilized to strengthen military positions in another area (e.g., the Soviet-Chinese border). In the second place, potential imbalances arising from the difficulty of equating different weapons systems would be excluded. (It would, for example, be difficult to say how many batteries of the ABM type would correspond to a cruiser, and so on.)

The next step in disarmament would entail proportional and simultaneous deescalation for all countries and in all strategic areas. Such a formula for "balanced" two-stage disarmament would ensure continuous security for all countries, an interrelated equilibrium between armed forces in areas where there is a potential danger of confrontation, while at the same time providing a radical solution to the economic and social problems that have arisen as a result of militarization. In the course of time a great many experts and politicians have put forward similar views, but hitherto these have not had significant impact. However, now that humanity is faced with a real threat of annihilation in the holocaust of nuclear explosion, I hope that we will not hesitate to take this step. Radical and balanced disarmament is in effect both necessary and possible, constituting an integral part of a manifold and complicated process for the solution of the menacing and urgent problems facing the world. The new phase in international relations which has been called détente, and which appears to have culminated with the Helsinki Conference, does in principle open up certain possibilities for a move in this direction.

The Final Act signed at the Helsinki Conference is particularly noteworthy because for the first time official expression was given to an approach which appears to be the only possible one for a solution of international security problems. This document contains far-reaching declarations on the relationship between international security and preservation of human rights, freedom of information, and freedom of movement. These rights are guaranteed by solemn obligations entered into by the participating nations. Obviously we cannot speak here of a guaranteed result, but we can speak of fresh possibilities that can only be realized as a result of long-term planned activities, in which the participating nations, and especially the democracies, maintain a unified and consistent attitude.

Regarding the problem of human rights, I should like to speak mainly of my own country. During the months since the Helsinki Conference there has been no real improvement in this direction. In fact there have been attempts on the part of hard-liners to "give the screw another turn," in international exchange of information, the freedom to choose the country in which one wishes to live, travel abroad for studies, work, or health reasons, as well as ordinary tourist travel. To illustrate my assertion, I should like to give you a few examples—chosen at random and without any attempt to provide a complete picture.

You all know, even better than I do, that children from Denmark can get on their bicycles and cycle off to the Adriatic. No one would ever suggest that they were "teenage spies." But Soviet children are not allowed to do this! I am sure you are familiar with analogous examples.

The UN General Assembly, influenced by the socialist states, has imposed restrictions on the use of satellites for international TV transmissions. Now that the Helsinki Conference has taken place, there is every reason to deal afresh with this problem. For millions of Soviet citizens this is an important and interesting matter.

In the Soviet Union there is a severe shortage of artificial limbs and similar aids for invalids. But no Soviet invalid, even though he may have received a formal invitation from a foreign organization, is allowed to travel abroad in response to such an invitation.

Soviet newsstands rarely offer non-Communist newspapers, and it is not possible to buy every issue of Communist periodicals. Even informative magazines like *Amerika* are in very short supply. They are on sale only at a small number of newsstands, and are immediately snapped up by eager buyers.

Any person wishing to emigrate from the Soviet Union must have a formal invitation from a close relative. For many this is an insoluble problem—for 300,000 Germans, for example, who wish to go to West Germany. (The emigration quota for Germans is 5,000 a year, which means that one might be forced to wait for 60 years!) The situation for those who wish to be reunited with relatives in Socialist countries is particularly tragic. There is no one to plead their case, and in such circumstances the arbitrary behavior of the authorities knows no bounds.

The freedom to travel and the freedom to choose where one wishes to work and live are still violated in the case of millions of collective-farm workers, and in the situation of hundreds of thousands of Crimean Tatars, who 30 years ago where cruelly and brutally deported from the Crimea and who to this day have been denied the right to return to their homeland.

The Helsinki Accord confirms the principle of freedom of conscience. However, a relentless struggle will have to be carried on if the provisions of this agreement are to be realized in practice. In the Soviet Union today many thousands of people are both judicially and extra-judicially persecuted for their convictions: for their religious faith and their desire to bring up their children in a religious spirit, or for reading and disseminating—often only to a few acquaintances—literature of which the state disapproves, but which from the standpoint of ordinary democratic practice is absolutely legitimate. On the moral plane, there is particular gravity in the persecution of persons who have defended other victims of unjust treatment, who have worked to publish and, in particular, to distribute information regarding both the persecution and trials of persons with deviant opinions and the conditions in places of imprisonment.

It is unbearable to consider that at the very moment we are gathered together in this hall on this festive occasion hundreds and thousands of prisoners of conscience are suffering from undernourishment, as the result of year-long hunger, of an almost total lack of proteins and vitamins in their diet, of a shortage of medicines (there is a ban on the sending of vitamins and medicines to inmates), and of overexertion. They shiver from cold, damp, and exhaustion in ill-lit dungeons, where they are forced to wage a ceaseless struggle for their human dignity and to maintain their convictions against the "indoctrination machine," in fact against the destruction of their souls. The special nature of the concentration-camp system is carefully concealed. The sufferings a handful have undergone, because they exposed the terrible conditions, provide the best proof of the truth of their allegations and accusations. Our concept of human dignity demands an immediate change in this system for all imprisoned persons, no matter how guilty they may be. But what about the sufferings of the innocent? Worst of all is the hell that exists in the special psychiatric clinics in Dnepropetrovsk, Sytchevka, Blagoveshchensk, Kazan, Chernyakhovsk, Orel, Leningrad, Tashkent . . .

There is no time for me today to describe in detail particular trials, or the fates of particular persons. There is a wealth of literature on this subject: may I draw your attention to the publications of Khronika Press in New York, which specializes in reprints of the Soviet *samizdat* periodical *The Chronicle of Current Events* and issues similar bulletins of current information. I should like to mention the names of some of the internees I know. I would ask you to remember that all prisoners of conscience and all political prisoners in my country share with me the honor of the Nobel Prize. Here are some of the names that are known to me:

Plyushch, Bukovsky, Gluzman, Moroz, Maria Semyonova, Nadezhda Svitlichnaya, Stefania Shabatura, Irina Stasiv-Kalinets, Irina Senik, Nijole Sadunaite, Anait Karapetian, Osipov, Kronid Lyubarsky, Shumuk, Vins, Rumachik, Khaustov, Superfin, Paulaitis, Simutis, Karavanskiy, Valery Marchenko, Shukhevich, Pavlenkov, Chernoglaz, Abankin, Suslenskiy, Meshener, Svitlichny, Safronov, Rode, Shakirov, Heifetz, Afanasiev, Ma-Khun, Butman, Lukianenko, Ogurtsov, Sergienko, Antoniuk, Lupynos, Ruban, Plakhotnyuk, Kovgar, Belov, Igrunov, Soldatov, Myattik, Kiirend, Jushkevich, Zdorovy, Tovmasian, Shakhverdian, Zagrobian, Airikian, Markosian, Arshakian, Mirauskas, Stus, Sverstiuk, Kandyba, Ubozhko, Romanyuk, Vorobyov, Gel, Pronyuk, Gladko, Malchevsky, Grazhis, Prishliak, Sapeliak, Kalinets, Suprei, Valdman, Demidov, Bernitchuk, Shovkovy, Gorbachov, Berchov, Turik, Zhukauskas, Bolonkin, Lsovoi, Petrov, Chekalin, Gorodetsky, Chornovil, Balakhonov, Bondar, Kalinichenko, Kolomin, Plumpa, Jaugelis, Fedoseyev, Osadchy, Budulak-Sharigin, Makarenko, Malkin, Shtern, Lazar Lyubarsky, Feldman, Roitburd, Shkolnik, Murzhenko, Fyodorov, Dymshits, Kuznetsov, Mendelevich, Altman, Penson, Knokh, Vulf Zalmanson, Izrail Zalmanson, and many, many others. Among those unjustly exiled are Anatoly Marchenko, Nashpits, and Tsitlyonok.

Mustafa Dzhemilev, Kovalev, and Tverdokhlebov are awaiting trial. There is no time to mention all the prisoners I know of, and there are many more whom I do not know, or of whom I have insufficient knowledge. But their names are all implicit in what I have to say, and I should like those whose names I have not announced to forgive me. Every single name, mentioned as well as unmentioned, represents a hard and heroic destiny, years of suffering, years of struggling for human dignity.

A final solution to persecutions can be based on international agreement—amnesty for political prisoners, for prisoners of conscience in prisons, internment camps, and psychiatric clinics as set forth in a UN General Assembly resolution. This proposal involves no intervention in the internal affairs of any country. It would apply to every state on the same basis—to the Soviet Union, to Indonesia, to Chile, to the Republic of South Africa, to Spain, to Brazil, and to every other country. Since the protection of human rights has been proclaimed in the United Nations Declaration of Human Rights, there can be no reason to call this issue a matter of purely internal or domestic concern. In order to achieve this goal, no efforts can be too great, however long the road may seem. And that the road is long was clearly shown during the recent session of the United Nations, in the course of which the United States moved a proposal for political amnesty, only to withdraw it after attempts had been made by a number of countries to expand the scope of the amnesty. I much regret what took place. A problem cannot be removed from circulation. I am profoundly convinced that it would be better to liberate a certain number of people—even though they might be guilty of some offense or other—than to keep thousands of innocent people locked up and exposed to torture.

Without losing sight of an overall solution of this kind, we must fight against injustice and the violation of human rights for every individual person separately. Much of our future depends on this.

In struggling to defend human rights we ought, I am convinced, first and foremost to protect the innocent victims of regimes installed in various countries, without demanding the destruction or total condemnation of these regimes. We need reform, not revolution. We need a flexible, pluralist, tolerant society, which selectively and experimentally can foster a free, undogmatic use of the experiences of all kinds of social systems. What is détente? What is rapprochement? We are concerned not with words, but with a willingness to create a better and more decent society, a better world order.

Thousands of years ago human tribes suffered great privations in the struggle to survive. It was then important not only to be able to handle a club, but also to possess the ability to think intelligently, to take care of the knowledge and experience garnered by the tribe, and to develop the links that would provide cooperation with other tribes. Today the human race is faced with a similar test. In infinite space many civilizations are bound to exist, among them societies that may be wiser and more "successful" than ours. I support the cosmological hypothesis which states that the development of the universe is repeated in its basic characteristics an infinite number of times. Further, other civilizations, including more "successful" ones, should exist an infinite number of times on the "preceding" and the "following" pages of the Book of the Universe. Yet we should not minimize our sacred endeavors in this world, where, like faint glimmers in the dark, we have emerged for a moment from the nothingness of dark unconsciousness into material existence. We must make good the demands of reason and create a life worthy of ourselves and of the goals we only dimly perceive.